THE COMPLETE CONSUMER'S GUIDE TO THE LATEST TELEPHONES

THE COMPLETE CONSUMER'S GUIDE TO THE LATEST TELEPHONES

Benedict A. Leerburger

Dodd, Mead & Company
NEW YORK

Published by Dodd, Mead & Company, Inc.
79 Madison Avenue, New York, N.Y. 10016

Distributed in Canada by
McClelland and Stewart Limited, Toronto

Manufactured in the United States of America

Designed by Berta Lewis

First Edition

Library of Congress Cataloging in Publication Data

Leerburger, Benedict A.
The complete consumer's guide to the latest telephones.

Includes index.
1. Telephone—Amateurs' manuals.
2. Telephone—Purchasing. I. Title.
TK9951.L44 1985 621.386'029 84-25839
ISBN 0-396-08506-7 (pbk.)

— CONTENTS —

INTRODUCTION

With the breakup of AT&T I realized that I could save a small fortune by buying and installing my own phones. No more monthly rental charges; no more old-fashioned equipment. I decided to go modern. I disconnected the old Ma Bell instrument that had served me faithfully and trouble-free for several decades and hauled it off to my neighborhood phone company store. In its place I plugged in a new, push-button instrument that looked more like a calculator than a phone. It was on sale at the local discount store—a real bargain. With the first incoming call my children broke out laughing. "That thing in the hall sounds like a sick bird, it chirps," said my oldest. To further deflate my ego, carrying on a conversation would have been easier using a tin can and a string. The sound quality on my end was bad, to be generous; it was near nonexistent on the caller's end. After a few moments of shouting I returned the compact instrument to the hall table only to see it slip to the floor and break into several pieces. Good old Ma Bell's phones never did that, I thought.

The next day I went to a local electronics store to replace my cheapie with a far more substantial instrument. I was immediately confronted with decisions: Did I want something called a "true touch-tone" push-button phone or something called a "pulse-dial" push-button? Did I want a ten-number memory feature, or would I prefer paying a few dollars less for a "simpler" phone and then buy an auto-dialer that could store up to 150 numbers? Did I want a cordless phone so I could receive calls in my back yard?

To be honest, I was too embarrassed to ask the salesman to explain the difference between the various push-button phones. I had never realized that pressing a phone's buttons did any more than connect me with the party of my choice. Further, I was too embarrassed to tell him that I didn't have 150 friends whose numbers could be implanted on his fancy memory device. I was beginning to have serious doubts about abandoning Ma Bell. Perhaps she would still take me back to the fold.

I did screw up my courage and ask the salesman if he had a book that a layperson could understand that explained what was going on in the world of telephones. I was sure others were having the same questions and problems trying to wade through the gulf left in the wake of poor Ma Bell's demise. To my surprise the salesman informed me that no such book existed but that he sure would welcome one because he had spent the entire morning answering the same questions I was asking for several other customers. I felt a great sense of relief that I was not unique in my ignorance. I also felt the challenge to solve the salesman's problem. Why not write the book to answer the very questions being raised by everyone interested in buying, installing, and servicing a new phone?

I called and met with dozens of dealers, manufacturers, and telephone distributors. Reading through manufacturers' literature and speaking with company engineers and marketing people were, of course, valuable learning experiences. However, actually using most of the phones described herein provided the best insight. As most consumers have discovered, what you are told and what you get are often quite different. One phone I looked at, for example, was called a "tone phone." There it was on the box in large letters. When I tried it out, however, it was a regular pulse-dialed instrument, not a true tone phone at all. As you'll discover in Chapter 3, the use of the word *tone* implies a very specific quality associated with telephones. This tone signal is essential if you use the long-lines services, transmit computerized data, use certain answering machines, and so forth. Without a true tone, or Touch-Tone as Ma Bell likes to call it, you have a standard, old-fashioned rotary-dial telephone. Even with push buttons, many phones work on the pulse, or rotary, system. I called the "tone phone's" manufacturer and asked how he could call his phone a "tone phone" when in reality it was pulse-dialed. Not to worry, I was informed; "Push the phone's buttons and you can hear a clear musical *tone*." So buyer beware!

The entire telephone business is in a state of confusion. Prices of both instruments and services are going to be drastically affected over the next several years. However, you'll have to tread with care. This book is a start. Read it and decide what kinds of phones and telephone services you want. Then ask your own questions. And most important, be sure you can test for yourself whatever instruments you want to buy. Don't forget: Compare, compare, and compare. You can get a good deal today.

Good luck!

For Julie, Marian, Ellen, and Christopher
who had to endure living with a grump
for too many months.

THE COMPLETE CONSUMER'S GUIDE TO THE LATEST TELEPHONES

—1—

The Breakup of Ma Bell

You want a red telephone in your new red and white kitchen. How do you go about finding just the right phone? Like shopping for a new refrigerator, you have a choice of many models. What shade of red do you want? Do you want a designer phone, a wall phone, push-button or dial? Do you want the telephone to record your most-used key phone numbers electronically? Do you want it complete with a paging system? Do you want it hooked up to a long-distance system? What brand of phone are you interested in? From what company do you wish to make your purchase? These are only a few of the myriad questions you might ask yourself before making your new telephone purchase. Before the breakup of American Telephone and Telegraph Company (AT&T), most of the choices were not yours to make. Ma Bell (the part-affectionate, part-derisive nickname for AT&T) controlled the telephone industry, making the phone system in the United States probably the best in the world—but also creating one of the world's giant monopolies. Finally in 1982, after a lengthy court battle, AT&T began to separate itself from its local companies and to dismantle this huge monopoly.

HOW AND WHY DID THIS BREAKUP OF MA BELL OCCUR?

Ever since the telephone company was begun by Alexander Graham Bell, who patented the telephone in 1876, Ma Bell was a firm disciplinarian to her many customers. She permitted no equipment to be used other than what she sold; no one else might tamper with her lines; no other company could tap into her long-distance system; and no other company could compete with her for services, telephones, or prices. Like a tolerant husband, the U.S. government not only allowed Ma Bell to control the industry but encouraged it for many years. The U.S. government granted it the exclusive right to provide the country with telephone service; Ma Bell's half of the bargain was to serve the public under government supervision.

And although she was a strict and even punitive parent, she took care of her young'uns. If your phone broke, she repaired it; if you couldn't find a telephone number, "Information" helped you; operators assisted you with long-distance calls. A popular diversion for years was to complain about the antics of Ma Bell, but if you had a phone problem, she always assisted you—when she got around to it!

Around the middle of the twentieth century, the government began muttering "monopoly" when complaining about AT&T. A landmark antitrust suit was brought in 1949, when the Justice Department argued that Western Electric, a division of AT&T, should be split off from the parent company because it did not permit other companies to sell phone equipment that would work with the Bell system's. That suit was settled by a consent decree in 1956 but probably was a key reason why another suit was filed against AT&T in the early 1970s. In 1956, while AT&T was permitted to remain intact, the consent decree insisted that AT&T pursue only the telephone business. Competitors now were licensed to develop telecommunications technology. The result was that AT&T was excluded from such areas as data processing and other fast-growing telecommunications changes.

In the 1960s and 1970s competitors taking advantage of the 1956 ruling began to challenge Ma Bell again. A major decision occurred in 1968, when the FCC ruled that if a piece of telephone equipment were mechanically compatible with the Bell system, AT&T could not refuse a customer permission to attach it. Shortly thereafter, in 1969, MCI Communications, with FCC permission, built a line to carry long-distance telephone calls between Chicago and St. Louis—and proposed rates much cheaper than AT&T's. A major threat to Ma Bell's dominance occurred in 1978, when a federal appeals court decreed that AT&T could not forbid competitors to link into its long-distance system, thereby challenging the giant corporation in the important market of long-distance phone service.

The 1970s saw several antitrust suits against AT&T, and when the Justice Department filed its own suit in 1974, arguing "abuse of monopoly," the mechanism was put into effect for the eventual breakup of the telephone company. Significantly, in 1976, U.S. District Judge Joseph Waddy ruled that the court rather than the FCC had jurisdiction over antitrust proceedings. A telling blow to Ma Bell occurred when the U.S. Supreme Court upheld this decision in 1977.

Begun in 1974, the antitrust suit went on for years and years. Ma Bell began to realize that she could no longer discipline her unruly children; they had grown up and had strength beyond adolescence. The officials of AT&T feared for their company, but meanwhile competitors were zooming ahead in various telecommunications areas from which Ma Bell was excluded by the 1956 consent decree. A key figure in the litigation was U.S. District Court Judge Harold H. Greene, who took over when Judge Waddy became ill. In 1981, to AT&T's chagrin, Judge Greene refused the company's request to drop the government suit. Recognizing clearly the inevitability of breakup, Ma Bell began plans to divest herself of her local operations companies.

In January 1982, the antitrust suit was officially settled, with the following ramifications for AT&T:

✆ First, by January 1, 1984, the company gave up control of its twenty-two local companies, which were regrouped into seven regional companies: U.S. West, Bell Atlantic, BellSouth, Pacific Telesis, NYNEX, Southwestern Bell, and Ameritech. Also known as regional holding companies, after another lengthy battle they were granted permission to use the Bell name and have a virtual monopoly on local telephone service. AT&T itself can no longer provide local telephone service to customers.

✆ Second, AT&T, now that it has ceased to be a giant monopoly, is no longer bound by the 1956 consent decree and can begin to explore telecommunications, data processing, and the computer fields. It has already formed a new subsidiary, AT&T Information Systems, which will research these areas. AT&T is permitted to hold Bell Laboratories, its research and development company, and Western Electric, its manufacturing division. Another major new subsidiary for AT&T is American Bell, Inc., which has two marketing divisions. The first, aimed at consumers, sells telephone equipment across the country in specially designated retail stores. The second division sells to businesses and specializes in automated office equipment. Its Dimension System 85, for example, is an office machine that coordinates telephone conversations with computer information. Sensitive to former labels of "monopoly," American Bell has specified that its Dimension System 85 will be compatible with equipment other than that manufactured by Bell so that other manufacturers' products may be attached to the system.

✆ Finally, AT&T must relinquish its monopoly on the long-distance telephone field. The company may still provide the long-distance services it always has, through AT&T

Long Lines, but the field is now open for multiple competitors to challenge AT&T's hold on this field. The courts have decreed that Ma Bell must allow competitors access to its local telephone connections.

HOW HAVE THE ANTITRUST PROCEEDINGS AND THE COURT'S RULINGS AFFECTED YOU, THE EAGER CONSUMER?

Oddly, you may not have noticed too many changes. You still make a phone call in the same way, you still have access to local and long-distance service, and you still receive a telephone bill each month that is larger than what you expected. But within these similarities exist huge differences.

For one, your telephone bills may have increased sharply in the last year. For many years, AT&T virtually subsidized local telephone service by, in effect, taking money from its long-distance division. The higher prices you used to pay for long-distance service kept your local rates low. Now that AT&T is no longer in the local telephone service business, you the consumer pay higher local rates to help the regional holding companies pay to maintain their lines. You also absorb some of the costs necessary for access to the long-distance lines.

A second major change for you is that you now have a choice of which long-distance system you want, and you will probably choose the one that provides the cheapest service for you. Competitors of AT&T Long Lines are numerous, and various packages are available to you. If you simply pick up your telephone and dial a long-distance number without an arrangement with a separate long-distance company, you are using AT&T Long Lines, which is now only one of the many long-distance services available to you. However, at some point in the future everyone in the country with a telephone may be forced to make a decision about which long-distance system he or she wishes to purchase.

Another annoying change is that you now may find that your telephone bills are much more complicated to translate. You may receive more than one bill, if you have chosen to use a long-distance service other than AT&T Long Lines. At any rate, your bill from the local company may include one section for local calls, one for long-distance calls, and one for equipment and service.

Amidst the confusion that followed the AT&T divestiture many new terms crept into the papers. Now, as the scramble to grab a piece of the communications pie is impacting on all consumers, a few definitions are in order:

Access Charges: Basically, this term refers to the rates long-distance carriers and end users pay the local telephone company for access to its facilities.

Answer Supervision: This is one means by which a switch recognizes when a call is answered. Answer supervision is a signal that is sent to the long-distance carrier's switch from the local telephone company when the called party answers the phone.

Automatic Number Identification: Local telephone switching center equipment identifies the telephone number of the calling party and sends this information to a message accounting/billing apparatus.

Central Office or End Office: This is the central switching unit of a telephone system where customer lines and trunks (multiple lines) terminate and are interconnected with the rest of the network. There may be more than one central office in an office building.

Consent Decree or Modified Final Judgment: On August 26, 1982, U.S. District Court Judge Harold Greene accepted, with modifications, the AT&T–Justice Department settlement terminating the government's 1974 antitrust suit against AT&T. His decree vacated the provisions of the 1956 consent decree that had kept AT&T out of competitive, unregulated ventures, and it required AT&T to divest twenty-two Bell operating companies.

Default Carrier/Default Traffic: If a local telephone subscriber doesn't designate a primary carrier for long-distance calls, the subscriber's calls may be automatically routed over the networks of one carrier or several carriers (on some proportional basis) at the direction of the local telephone company.

Divestiture: AT&T was required to spin off its twenty-two local operating companies as a condition for ending the government's 1974 antitrust suit (see also *Consent Decree*).

Equal Access: As required by the settlement entered in the AT&T antitrust suit, all long-distance carriers will be offered the same local access to lines to connect to local subscribers now provided generally only to AT&T. Previously, non-AT&T carriers connected with their local subscribers at the central office through ordinary business lines. AT&T, and now the other carriers, enjoy a "trunk-side" connection that provides access to all local subscribers, including those with rotary-dial telephones and, among other things, identifies the calling party, registers when the called phone answers, and generates billing information. Equal access officially began September 1, 1984, and is to be substantially completed by the Bell System by September 1, 1986, when all central offices serving ten thousand lines or more are to convert their switching facilities so they can offer all long-distance carriers the same local connections.

Exchange Network Facilities for Interstate Access (ENFIA): A negotiated agreement under which other common carriers, since April 1979, have compensated local telephone companies for local access lines to connect with their subscribers.

Other Common Carriers (OCCs): Carriers providing interexchange communications services in competition with AT&T.

Presubscription: Local telephone subscribers in equal access areas will be asked to choose a long-distance carrier to automatically handle their long-distance calls.

Primary Carrier: The carrier designated by local subscribers through presubscription.

There is little doubt that the impact of AT&T's divestiture will be felt by all consumers. Decisions must be made. Finally, as a result of Ma Bell's breakup, you have a vast choice of equipment from which to choose. No longer are you limited to AT&T's selection of telephones, colors, models, and gadgets. Dozens of equipment manufacturers have entered the field, offering you a dizzying array of merchandise. Telephones can look stylish, "do tricks," and serve you in ways you might never have thought possible.

–2–

Profiting from AT&T's Breakup

Prior to the AT&T breakup, any attempt to alter your phone system without the knowledge of Ma Bell usually brought nasty letters and threats of legal action. The phone company jealously guarded its monopoly. However, even as the courts were thrashing about AT&T's future, a slew of telephone equipment manufacturers, marketers, and purveyors of home communications supplies began flooding the market with their wares.

Within a matter of months following the January 1984 date on which Ma Bell officially became eviscerated, more than two dozen companies entered the home equipment market. Today customers are offered a wide variety of telephones ranging from Mickey Mouse and Kermit the Frog phones to sophisticated devices that do everything from answering the front door and monitoring your home's smoke level to actually allowing you to speak across the country the old-fashioned way. Never before has the American buying public been offered such an array of gadgets, gewgaws, and gimmicks for home communications. Fortunately, amidst the options open to the serious buyer of home telephone devices, there is a wide range of well-designed, useful, and surprisingly inexpensive equipment. Considering the inflationary trend, who would have assumed a few years ago that you could stop into your local department store and buy a usable telephone for under twenty-five dollars?

Unfortunately for the consumer, however, the number, variety, and quality of products is mind-boggling. Touch-tone, pulse-dial, automatic redial, auto-dial, rotary, push-button—the terms are endless. This book will attempt to plow through the maze and provide a better understanding of the problems, products, and pitfalls that may await.

First, a few simple "cans" and "cannots" based on the Federal government's ruling that led to the breakup of AT&T:

You now can:

1. *Buy your own telephone and telecommunications equipment.* Several years before the government decided that Ma Bell was operating a monopoly, the courts ruled that it is legal to own your own telephones. Although the phone company didn't like competing in the marketplace, it really didn't have much choice. Ma Bell still designed and built its hardware through its Western Electric subsidiary and did its best to discourage the use of competing equipment. The phone company made it quite clear that if you bought a competing product and installed it yourself, you were fully responsible for any equipment problems. Of course, if you bought the company's equipment, Ma Bell would back it to the hilt with unlimited free service.

AT&T still sells a wide range of telephones and telephone appliances to the consumer through its phone stores. And as we'll discover in a later chapter, AT&T distributes a high-quality product. Further, its range of consumer phone products is wide enough to satisfy just about every taste. However, buying an AT&T phone doesn't give you anything but an instrument. You are still responsible for maintaining your phone as well as all the ancillary equipment you choose. AT&T, as well as many other equipment suppliers, offers a warranty on its products. However, the burden of maintenance and repair is up to you. Obviously, the type of equipment you elect to buy is up to you. Your choice is wide and growing wider. In most cases the various accessory options are compatible with your basic phones. Thus, you are at liberty to buy several phones manufactured by several manufacturers and couple them with auto-dialers, an answering machine, and an amplifier—all made by different companies. There are some limitations based on your telephone wires, but for the majority of basic home hookups, you're free to piece together just about any type of home communications network you desire.

2. *Use the current telephone wires in your home.* When you pay your monthly bill to your new regional phone company, you pay for the use of the telephone lines leading to and from your home as well as the telephone wires within your house. (If you don't buy your own phones, you also pay a separate rental fee for the use of the phone company's hardware.) Although you are free to add your own phones to the telephone company's wires, the Federal Communications Commission requires you to contact your local phone company and let it know how many new phones you add as well as the brand names, model numbers, registration numbers, and the ringer equivalence numbers. The latter two names may sound confusing, but they are actually easy-to-find numerals usually located on the phone's base.

In most cases the present array of phone wires will not be properly located for any major phone add-ons. Thus, you'll have to do some of your own wiring to and from the present lines leading into the small square "junction boxes" along your baseboards. In most states you are allowed to do your own internal wiring. In some states, however, it is necessary to contact your local phone company to do the wiring—at your expense, of course.

Obviously, new phone wires can only be connected to other phone wires. In *no* case can phone wires be connected to electric wires or cords. Wiring your own phones is a relatively simple and safe operation. Once you start mixing phone lines with the electrical wires installed by your local utility company, however, you can find yourself in serious trouble. The simple solution: stay away from electric wires.

3. *Use the long-distance service of your choice.* With the AT&T breakup came a group of companies anxious to sell you a "better, cheaper" long-distance service. Since Ma Bell's long-distance rates were regulated by the Federal Communications Commission as well as by many state and local regulatory agencies, she became an easy target for price choppers. With the increased competition, the consumer has benefited. The cost of long-distance service has been reduced and by careful shopping for a long-distance service it is significantly cheaper to place a long-distance call. AT&T, still very much in the long-distance business, is countering the competition by price cutting and trying to offer better service. For example, Ma Bell claims that only the phone company allows you to talk directly to another human being if you have a problem reaching your party. And have you noticed how polite the operators are lately? Further, only AT&T offers specialized services such as person-to-person calls, collect calls, and credit card calls. The customer, however, does have a choice: AT&T or the competing services.

4. *Hook your personal computer into the phone lines.* Through the use of a device called a "modem" any personal computer can be programmed to communicate electronically with any other similarly equipped computer. A simple connection is made to any phone outlet. The advantages of using a computer to communicate long distance are many. Electronic mail, drawings, graphs, charts, and data are delivered at high speed, and thus lower cost, between individuals using their personal computers. When you rent the lines from your local phone company you can use those lines for either oral or electronic messages.

You cannot:

1. *Tie directly into the telephone lines leading into your home.* These wires belong to your regional phone company. If you want additional wiring installed into your home, you must still deal with the phone company.

2. *Hook into a pay phone.* Although it sounds obvious, shortly after the breakup AT&T received several letters asking if the local phone box was now fair game for do-it-yourselfers. It's not!

3. *Connect a cordless phone to a party line.* If you have a party line and want to use a cordless phone, you need the phone company's permission. They will probably install a protective coupler at your expense that will allow you to override the party line. You may also be billed monthly for this added equipment, depending on the policy and generosity of your phone company.

4. *Remove a phone company's equipment from your home.* If you pay a monthly fee to your local phone company to rent your equipment, the telephones and any other ancillary devices still belong to AT&T. In some cases a local phone company will allow you to take your rented equipment with you if you move within their area of phone coverage. In all cases you first must receive company approval.

OWN OR RENT?

Basically, you have already made your initial decision. If you didn't elect to buy the phones in your home when the phone company first contacted you with the option of buying your present equipment at a discount rate, you are renting your phones from AT&T, not your local phone company. There is certainly nothing wrong with a decision to hold on to your present phones and wait and see what happens in the telephone marketplace. Many people don't want the bother and confusion that go with buying and installing their own equipment. Rather, they prefer to keep the status quo. With each month's bill comes a small charge for equipment. Each of the various regional phone companies spells out quite clearly what each instrument costs per month to rent. Your new regional phone company acts as an agent for AT&T in collecting a rental fee for the equipment you elect to lease. If there is any problem with your phone, then it's AT&T's problem. After all, it is their phone; they should take care of repairing or replacing it. And AT&T agrees with this approach. They love renting phones. And if a problem does occur, then AT&T will take care of it. In most cases phone problems are not a result of faulty instruments but of trouble "down the line." This trouble can often be cleared up within a matter of minutes at the company's switchboard.

If there is a problem with a broken or faulty wire anywhere from the company's switchboard to your house, the phone company will fix it. Of course, even if you own your phones, the phone company will still fix any wire problem leading to your house. However, the family that decides to rent AT&T's equipment also has the assurance that wire problems within the home are problems that will be handled by the phone company. For the phone owner it's another story. Any internal wiring problem is now your problem. It should be mentioned that problems within the wires are extremely rare. If you're going to have a problem, it is usually in a do-it-yourself connection or a piece of equipment. The internal wires themselves are the last place to expect trouble. Problems with the telephones are another matter. If you own your own phone, you are responsible for repairs. Even if your equipment is covered by a warranty, you still must find a local or regional repair station, take or ship your phone—and then wait. Depending on the phone's manufacturer, you can wait a few days or quite a few weeks. You also pay for shipping the phone back to the manufacturer and, in some cases, a "small" service charge ranging anywhere from two or three to over ten dollars.

Consider that a phone can cost as little as twenty-five dollars; but you may end up paying as much in postage and service charges—not to mention inconvenience—as a new, higher-quality telephone costs. For many, therefore, the home telephone has joined the ranks of "disposable" items along with razors, pens, and plastic cups.

On the other hand, what does it cost to rent a phone from AT&T? And how does the rental cost compare with the cost of buying a new phone? AT&T has standardized most telephone rental costs throughout much of the country. The following desk phones, for example*:

Standard Rotary-Dial Phone	$1.50 per month
Standard Touch-Tone Phone	$2.85 per month
Standard Rotary Trimline Phone	$3.45 per month
Standard Touch-Tone Trimline Phone	$4.60 per month
Princess Rotary-Dial Phone	$3.15 per month
Princess Touch-Tone Phone	$4.05 per month

In addition to your monthly rental fee you may also be taxed by your local state, city, or county government. It is a good idea, however, to check with your local phone company for its exact charges. Now let's compare the cost of renting a phone with the cost of buying the identical phone new from AT&T. Although AT&T doesn't sell all the phones it leases, it carries most of them and has a wide range of additional phones from which to select. More about these options in a later chapter.

AT&T will sell you a standard rotary-dial desk phone for $42.95. (Its standard Touch-Tone phone is ten dollars more; its Trimline Touch-Tone desk phone is $67.95.) Your total rental charges for a standard rotary-dial phone will match the cost of buying an identical phone in twenty-nine

* *Note:* These rates apply, as of September 1984, to all states except Arizona, California, Colorado, Delaware, Idaho, Louisiana, Maine, Michigan, New Jersey, Oregon, Washington, Wyoming, and Nevada, where there is a slight variance.

months (assuming the cost of state and local taxes equals the loss of potential interest on the money invested in the original telephone plus depreciation). Your pay-back on a standard Touch-Tone phone is nineteen months, and on a Touch-Tone Trimline desk phone, seventeen months.

These comparisons are based on comparing identical telephone equipment. Once you start shopping for phones that are less expensive than the AT&T phones, your pay-back drops off sharply. For example, ITT manufactures a desk model tone phone that can be purchased for about thirty-five dollars. The rent vs. own pay-back drops to thirteen months. With some of the less-expensive equipment available at discount stores, you can easily reduce the pay-back period to under a year.

Another factor to consider before buying your own telephone equipment is the obsolescence factor. Each year at the Consumer Electronics trade show, manufacturers and telephone-product vendors display their newest equipment. Some manufacturers introduce new products at least twice a year. Although much of the new equipment is but floss added to the old to provide a shiny new look, with the rapid increase in technical developments there are often new features added to old-line phones, or new products introduced that are indeed worthy of consideration.

When you lease AT&T's equipment you always have the option of upgrading your hardware for improved or newer devices. AT&T is limited in what it markets, but with the technical know-how of its Bell Laboratories subsidiary behind it, you can be sure it will remain in the forefront of telephone technology. Owning your own equipment locks you in. You can always upgrade your phones and equipment for newer models, but by doing so you must write off the cost of your present hardware. There isn't a market, as yet, for used consumer telephones.

If you're confused by the maze of products, services, and options now available—wait a minute; there will be more tomorrow! The home communications business is BIG BUSINESS and no company worth its stockholders is going to let a potential multibillion-dollar market slip by without taking a shot at its share.

The drive to sell hardware is nothing compared to the business of selling services—particularly long-distance services. After all, when you buy your own telephone you pay only once. You pay for long-distance calls every month. And the long-distance market generates more than $45 billion in calls within the United States alone. You can be sure that Ma Bell isn't going to give up a dime of that market without a fight. Unfortunately for the consumer, the struggle among the telephone long-lines giants means more confusion. Amidst the confusion, however, will the competitive battle mean cheaper calls? Only time will answer this question.

— 3 —

Selecting a Long-Distance Service

Before the end of 1986 a new era in telephone service will begin. You will be asked to make a decision: Which of the several competing long-distance services is to be your primary long-distance carrier? Your choice: AT&T, Sprint, MCI, SBS Skyline, ITT, Western Union, or one of the host of smaller regional networks that are geared up for a long, bloody, competitive battle. The reason behind selecting a primary long-distance carrier originated in the 1982 ruling that sent Ma Bell's corporate children off on their own. As part of the split-up U.S. District Judge Harold Greene ruled that each separate operating telephone company had to make it just as easy for a customer to call long-distance as AT&T. The so-called "equal access" provision was intended to promote competition in long-distance service by preventing the existence of a monopoly and thus to keep costs under control. What has resulted is a fierce battle among a few giants with a slew of Davids slashing away at everything in sight.

In the summer of 1984, Charleston, South Carolina, soon followed by Alameda, California, became the first American cities technically hooked up and ready to offer phone customers their choice. Prior to that date, customers wanting to buy a service other than Ma Bell's had to be located in a region in which long-distance service companies other than AT&T operated. And they had to have a true tone or modified dialing system. The musical tones of true touch-tone are necessary because the long-distance services other than AT&T require the electronic signals generated by each specific musical tone to activate various switching devices. Since only about 60 percent of all telephones in America have the true touch-tone capability, Ma Bell's competitors are most anxious to fight on equal grounds.

An additional factor that the courts have changed is the multitude of buttons that you have to press to make a long-distance call with any one of Ma Bell's competitors. As you know, to call long distance without a special long-line service requires, in most areas, dialing the numeral 1, followed by a three-digit area code and your party's seven-digit number: a total of eleven numbers. Prior to equal access, using the long-lines service companies required dialing a seven-digit local number to hook you into the service. After waiting for the unfriendly computer tone, another six-digit number is necessary to identify yourself and assure the company that it has a legitimate party to bill. Finally, you dial the area code (three digits) and the number of the party you want to reach (another seven digits). You have now dialed a total of twenty-three numbers. Get a busy signal, and it's back to square one. God forbid you err on one of those last ten numbers. You can wind up with a wrong number three thousand miles away. (One of the phone company's competitive claims is that it will give you credit if you get a wrong number simply by dialing 211 and telling an operator your problem. "Try that with the long-lines service companies," it argues. And prior to equal access it was right.)

During the hot summer of 1984 the competitive battle was heated still further when Ma Bell announced that she was taking a trick from the airlines, credit card companies, and banks by giving goodies to nonbusiness customers who are solid AT&T users. The program, dubbed "AT&T Opportunity Calling," applies to anyone running up a monthly long-distance bill of fifteen dollars or more. The caller is now eligible to receive coupons good for discounts on everything from radial tires, microwave ovens, and airline tickets to a room for the night at a Howard Johnson's motel. Thus, if you run up a monthly long-distance bill of fifty dollars, AT&T will give you a fifty-dollar credit toward the acquisition of a host of products or services. AT&T provides a catalog illustrating the variety of goods available. The discounts are limited to $500.00 and apply only to domestic calls.

Following the AT&T announcement, MCI countered by cutting its prices by up to 6 percent and proudly boasting that "AT&T has apparently conceded the price issue." Not to be outdone, Sprint tried to turn AT&T's advertising claim of "better, more personal service with operator-assisted calls" around by arguing that Sprint saved its customers money by cutting out the middleman.

HOW LONG-DISTANCE CALLING WORKS

When Alexander Graham Bell spilled acid in his laboratory and shouted, "Come here, Watson! I need you," his voice was carried to his lab assistant in the adjoining room over a wire. From that moment until the mid-twentieth century, the telephone relied on a thin strand of wire to carry vocal communications. In the 1950s more exotic forms of technical innovation proved more efficient than wires. Thin strands of flexible glass were developed that carry almost ten times as much data as a similar size copper wire. Fiber optics is but one of many technical developments that are rapidly finding their way into our lives.

Have you noticed those dish-shaped objects atop buildings, steel towers, and even on the peaks of mountains? These microwave antennae serve both as electronic-signal-transmitting stations and also as long-distance relay stations.

Like the old children's game Pass-it-On, microwave signals are transmitted from one dish to another along a crazy-quilt network that blankets the hemisphere. If the high traffic of signal transmission causes congestion along one route, high-speed computers determine a faster way to transmit a signal. For example, you can call a friend in California from your home in New York and have your signal zigzag along a microwave network from Boston to Chicago to Dallas to Los Angeles while your friend's signal returns to you via microwave by a completely different route.

If the computers find an overload on one microwave circuit, your call may skip the microwave route and be transmitted long distance by bouncing your signal off an orbiting satellite. Overseas calls are almost all transmitted via satellite. One long-lines service, SBS Skyline, relies almost exclusively on satellite transmission.

The long-distance telephone network in the United States is the most advanced, most sophisticated of any in the world. Considering the fact that millions of Americans are out of work, it is amazing that we own just about one phone per person. There are more telephone poles in America than people living in the state of New York—20 million poles, to be exact. Is it any wonder that the competition to get into the long-lines communications business is going to be a battle of giants?

AT&T Long-Distance Calling

Whether you call her Ma Bell or AT&T, the old gal is still the dominant force in world communications. The company that has pioneered every major development in the communications field, whose scientists and engineers have won more Nobel Prizes than any other corporation, the company responsible for the invention of the transistor, isn't going to lie down for anybody. To counter the competition in the summer of 1984, AT&T announced its "Reach Out America" program. Here's how it works.

For a flat rate of ten dollars a month you receive one hour of direct-dialed weekend and night long-distance calling time. (Weekends run from 5:00 P.M. Friday until 8:00 A.M. Monday; nights run from 11:00 P.M. until 8:00 A.M. the following morning.) Each additional hour costs $8.75 (prorated if the additional time is less than one hour). Want an additional 15 percent off evening long-distance rates? (Evening is 5:00 P.M. until 11:00 P.M.) Just add another $1.50 per month. AT&T imposes a one-time connection fee of ten dollars if you decide to join the plan. No special phone line or Touch-Tone is necessary. Just call your local AT&T service representative or 800–222–2300.

The idea of charging solely on time rather than on distance plus time is a leaf from SBS Skyline's book. How do AT&T's "Reach Out America" rates compare with Skyline's? At ten dollars per month for an hour of late-night/weekend calling times, you pay $0.17 per minute compared with Skyline's charges of $0.10 per minute to call a neighboring state and $0.14 per minute to call all other states. That gives Skyline a 17 percent cost advantage over AT&T for any calls other than those to states next door. Make a call over the state line, and Skyline's cost advantage is upped by 41 percent. However, AT&T boasts that it can do things no other long-lines service has yet been able to achieve. For example, it offers:

- The ability to call from anywhere to any location in the world at any time. This includes direct dialing to 103 countries and operator assistance to more than 250 countries.
- Immediate credit for misdialed or disconnected calls.
- Operators to help you, available twenty-four hours a day.
- Collect calling.
- Person-to-person calling.
- A personal AT&T credit card.

AT&T further boasts more than forty thousand operators to help its customers. This is a major plus in the long-lines war. But who really pays the salary for these personnel? Compare AT&T's rates with those of its competition, and you'll find the answer. Ma Bell also claims that "with more than 54 million miles of circuits and 54 million billion possible connections, you can be sure that virtually every long distance call you make will be completed promptly in one attempt, even during the busiest calling time of day." For the customer who must decide between long-lines service companies, AT&T is striving to stake out its image as the Cadillac of companies. Not a bad idea if its claims to being the most efficient of telecommunications companies holds up. I have tried the competing services and agree that there have been problems with all. You can receive a poor connection, particularly if a satellite transmission goes awry. And with AT&T's principal competitors, if you reach a wrong number you won't have an easy time trying to reach a credit operator. However, the customer must weigh the pros against the cons. You do get what you pay for. Get a wrong number with MCI, Sprint, SBS Skyline, or one of the others, and you are out anywhere from thirteen to seventy-five cents, depending on the time of day and the distance of your call. You must decide how many wrong numbers you reach and weigh the cost factors. Make no bones about it, AT&T is good and it is going to get better, better not just in terms of service but also in terms of cost efficiency. Don't be afraid to switch from Ma Bell to another service; but don't make it a farewell kiss. You may be back! If you don't want its "Reach Out America" program, what about AT&T's standard rates?

In addition to the familiar AT&T, let's take a look at the other major long-lines companies that are after your dollar.

MCI Communications

Contact: 800–243–2140

It was MCI that rocked Ma Bell with the civil antitrust suit in 1974 that charged AT&T and the Bell operating

*AT&T Per-Minute Rate Range**

WEEKDAYS 8:00 A.M. to 5:00 P.M.	$0.16 to $0.69 for the first minute
WEEKNIGHTS 5:00 P.M. to 11:00 P.M.	$0.10 to $0.41 for the first minute
LATE NIGHT 11:00 P.M. to 8:00 A.M. 5:00 P.M. Friday to 5:00 P.M. Sunday	$0.06 to $0.27 for the first minute

*AT&T bases its charges on the formula: time plus distance; there is a specific rate for the first minute or portion of a minute, plus a lower rate for each minute or portion of a minute that follows. For the exact rate from your calling city to any city in the world, contact your local phone company.

companies with monopolization in violation of the Sherman Antitrust Act.

This is one of the new giants in the long-lines service battle. MCI was founded less than two decades ago, in 1968, and already it has built its business to serve more than 2 million customers. In 1984 its annual sales exceeded $1.6 billion. MCI has spread quickly to grab a foothold in four prime areas of the communications business: domestic long-distance service; mobil and cellular telephones; overseas message and data service; and its newest, digital information services—electronic mail. For the home telephone consumer, however, MCI is an interstate long-lines service company.

As a long-lines company, MCI provides service to all the continental states plus Hawaii, Puerto Rico, the Virgin Islands, and parts of Canada. It is also authorized to provide intrastate long-distance service to eleven states: Alabama, California, Florida, Indiana, Iowa, Maryland, New York,

Ohio, Pennsylvania, Texas, and West Virginia. It has its own twenty-thousand-mile coast-to-coast terrestrial microwave network and is installing another 4,200 miles of optical fiber to increase its network capability. MCI also uses both satellite facilities and rented lines to transmit more than 6 billion minutes of domestic long-distance traffic each year. It is behind only the titanic AT&T network as a provider of domestic long-distance services.

The various MCI services include:

✆ *MCI Standard Service:* In areas where equal access has not yet taken hold, MCI offers its Standard Service. Subscribers receive a personal code and can call anytime from a true tone push-button phone as long as they are in an area that has an MCI local access number. There are no subscription fees, installation charges, or monthly minimums. If you would like an "optional calling card," which allows you to use the MCI service from any company access phone number throughout the country, it's yours for the asking at no extra charge. Rates are based on distance plus time to the nearest minute. And of course MCI promises to save you money compared to Ma Bell's rates.

✆ *MCI Dial "1" Long Distance Service:* This service, inaugurated in Charleston, West Virginia, in the summer of 1984, provides local subscribers with the same access to long-distance service as AT&T. Subscribers can call anytime using either a rotary or a true tone push-button phone simply by dialing "1" plus the area code and number. Their calls will be automatically routed over MCI. There is no subscription fee, installation charge, or monthly minimum usage requirement. Rates are based on distance plus time. The company claims it will be saving its customers "up to 40 percent over AT&T." Volume discounts are also provided. Although the service is in limited use throughout the country, as equal access becomes effective so will MCI's Dial "1" Long Distance Service.

Sprint

Contact: 800–521–4949

In 1970, one year after MCI decided to go after Ma Bell's monopoly on long-distance service, the Southern Pacific Railroad decided to expand its own private microwave telecommunications network that it used to control its business along the vast right-of-way between California and the Southwest. It formed a subsidiary company, Southern Pacific Communications (SPC), to develop a coast-to-coast private-line communications service. When, in 1977, federal court decisions confirmed the right of specialized common carriers to provide long-distance telephone services similar to those provided by AT&T, SPC was off and running.

In late 1978, SPC introduced its Sprint long-distance telephone service with sixteen thousand customers. Within five years Sprint had enlisted almost one million customers and had linked all metropolitan areas coast to coast. In 1983, SPC was acquired by GTE and became GTE Sprint Communications. Within months of the acquisition, Sprint got a jump on its competitors by announcing that the Sprint network could reach any telephone within the fifty states, Puerto Rico, and the U.S. Virgin Islands. Prior to this, only AT&T could boast that it could make the right connections between all phones in America. In fact, having direct access to only the metropolitan areas was great for businesses, but it put a damper on consumers who wanted to call beyond the big cities.

Sprint is a rapidly growing long-lines service company with a network extending some 46.2 million circuit miles. It is geared to the competitive battle and is prepared to slash prices to meet the competition.

Sprint's pricing is based solely on distance, duration, and time of day of the call. Some competing services have a flat, unit cost per call. Basing costs on distance alone may make it cheaper, in comparison with the flat-cost method, to call neighboring states but more expensive to call long distances.

There is no monthly service charge, and no additional equipment, other than a true tone phone, is required.

Some of Sprint's services include:

✆ *Volume Discounts:* For large-volume calling, discounts are offered over and above its regular rates on interstate calls. Once a customer's bill reaches $25.00, a discount is automatically applied to that first $25.00 worth of calls and on every call thereafter. Greater discounts are given when the bill hits $75.00 and again at the $200.00 level.

✆ *Travelcode:* This feature allows Sprint customers to reach an access number from any Sprint network city. By allowing customers to call from any major city, the company hopes to counter AT&T's claim that only they provide credit card and collect call service.

✆ *Speedial:* If you frequently call one number, Sprint says it will make an arrangement whereby that ten-digit number (including area code) is replaced by a special four-digit code that simplifies the dialing process. Of course, for the customer who lives by the phone, any of the automatic dialers will accomplish the same result. However, Sprint says it will give its good customers an automatic dialer called The Sprint Edge, which will save you making the investment.

Satellite Business Systems (SBS Skyline)
Contact: 800–232–6400

When the first rumors of the demise of AT&T's monopoly hit the boardrooms of America's top corporations, several high-powered execs realized that the biggest users of long-distance services were themselves—the nation's businessmen. Why not set up a private long-distance network for the business community? With this germ of an idea leaders from IBM, Aetna Life & Casualty, and Communications Satellite Corporation decided to form a company that could use the computer technology of IBM, the satellite and communications expertise of ComSat, and the managing and marketing ability of Aetna to serve the needs of both the residential community and small businesses.

By the end of 1984 SBS served well over a hundred thousand customers. And why not? Skyline offered a variety of extras that kept it a few steps ahead of its larger competitors. For example:

✆ *Rates based on distance not the key factor.* In January 1985, SBS abandoned their two-tier price structure (neighboring and non-neighboring states) to add a third level: calls between east and west coast states. Their current rate structure is thus: Calls between adjacent states; calls to non-neighboring (SBS calls them "regional") states; and coast-to-coast calls. Thus, for the customer living in New York, it costs no more to call Texas than it does New Hampshire. Since Skyline is bouncing your telephone signal off a passing satellite, it figures it is just as easy to reflect the return signal to Texas as it is to bounce it back to a closer area. And it can save on line rental by using its own three-satellite network. (Skyline sent its third satellite into orbit as part of the Space Shuttle's initial commercial payload.)

✆ *No monthly subscription charge.* Although Skyline is no longer unique in not charging customers a monthly fee to use the service, it was the first. Sprint and MCI dropped their monthly service charges when the competition started to heat up. Skyline does impose a monthly minimum rate of fifteen dollars. However, if you don't spend more than fifteen dollars per month on long-distance calls, you should reevaluate your need for the service.

✆ *No local message-unit charges.* In some parts of the country your local phone company imposes a unit charge on local calls. Skyline supplies its customers with a local number for which no unit charge is imposed.

✆ *Fractional minute rates.* Most long-distance services, including AT&T, Sprint, MCI, and Western Union's Long Distance Services, charge on a per-minute basis. If your call lasts 2 minutes and 12 seconds, you are charged for a 3-minute call. Skyline breaks your call into thirty-second segments following the first minute. Thus, your 2-minute , 12-second call is billed as 2.5, not 3 minutes. This feature may not sound like much, but over a period of time it will save you quite a bit. Considering that a 2-minute, 1-second call is billed as a 3-minute call by many long-distance services, this is a major plus for Skyline. Skyline also starts the clock running when your party answers their phone. Don't be surprised! Some long-lines companies start the clock when the phone rings, and you find yourself charged for unanswered calls.

✆ *Discounts for all customers.* Pay your Skyline bill before the "discount due date" stated on your invoice, and the company gives you a 2 percent discount, deducted from your next bill. You have about ten days to two weeks following receipt of the bill to take advantage of this cost-saving benefit.

✆ *Use of the service away from home.* If you travel to other cities or wish to give your relatives access to your private authorization code, Skyline will allow you to call its local access number in other cities along its network. This is not a free service. You are charged four dollars per city.

Until midsummer 1984, Skyline maintained a sixteen-dollar installation fee but dropped it when it felt the heat of its competitors, who had eliminated the charges to hook you into their systems. Although Skyline can reach every phone in the country, it has not, as yet, fully geared up to allow every phone to make outgoing calls. It has the major metropolitan areas tied in and it is currently developing access numbers for the rest of the country.

SBS Skyline Rate Schedule

WEEKDAYS		
8:00 A.M. to 5:00 P.M.	Neighboring State	$0.27 per minute
	Regional	0.37 per minute
WEEKNIGHTS		
5:00 P.M. to 11:00 P.M.	Neighboring State	$0.15 per minute
	Regional	0.20 per minute
	Coast to Coast	0.24 per minute
LATE NIGHT		
11:00 P.M. to 8:00 A.M.	Neighboring State	$0.11 per minute
	Regional	0.15 per minute
	Coast to coast	0.17 per minute
WEEKENDS		
5:00 P.M. Friday to 8:00 A.M. Monday	Neighboring State	$0.11 per minute
	Regional	0.15 per minute
	Coast to coast	0.17 per minute

ITT's "Longer Distance Service"
Contact: 800–526–7270

International Telephone & Telegraph (ITT) is an old-line company that has been a pioneer in the communications business for many decades. It was one of the first companies to build its own system, a network that now blankets most of the East, South, and Southwest. ITT concentrates on serving the business community, but for the consumer who is a heavy phone user it is worth investigating. Its products include a private line service designed to provide high-speed communications point to point, a link-up for the hotel industry, a "cost-effective WATS alternative" that ties the heavy phone user into ITT's national network, and its "Longer Distance Service," which uses the ITT network in those metropolitan areas covered. It leases AT&T lines in non-covered parts of the country. Its advantages include:

✆ *Multiple access capacity.* More than one phone can use the service at the same time.

✆ *Rates lower than AT&T's.* However, with cut-throat competition it is hard to predict who will have the lowest rates tomorrow.

✆ *Discounts available to heavy users.*

ITT charges a monthly fee to use the service as well as a start-up charge. Its rate schedule is based on two factors: calls from one ITT "Nationwide" city to any phone in the continental United States, and calls from one ITT "Nationwide" city to any other locale on the ITT network. The latter offer rates about 10 percent less than the open-ended program.

Its basic rate schedule from any phone in an ITT network city to any phone in the country is:

	Per-Minute Rate Range
WEEKDAYS	
8:00 A.M. to 5:00 P.M.	$0.16 to $0.513 per minute
WEEKNIGHTS	
5:00 P.M. to 11:00 P.M.	$0.091 to $0.292 per minute
LATE NIGHT	
11:00 P.M. to 8:00 A.M.	$0.059 to $0.19 per minute

Western Union Long Distance Services
Contact: 800–562–0240

There's been quite a change since Thomas Edison received his first job sending Morse code messages on a telegraph key for Western Union. Today, Western Union is firmly entrenched in the telecommunications field and is a growing force in the battle to get your long-distance business. Although Western Union is making a major push to enlist the corporate community, it is smart enough to understand the value of the consumer market. For the business market it has established such features as allowing separate accounting codes to be assigned to each department or even each employee. Up to ninety-nine separate codes can be assigned. If you have a large family and make a lot of long-distance calls, this feature may be worth thinking about. You can also use its "Abbreviated Dialing" feature to cut back on the number of digits you have to punch. After dialing your six-digit authorization code you can select up to sixty of your most frequently dialed numbers and have a separate three-digit code number assigned to each.

Calls can be placed anywhere in the United States any time of day or night. Travelers can rent a separate access code number for three dollars per month that will allow on-the-road calls in any of Western Union's 150 U.S. network cities. There is no charge to be connected to the system. Charges are based on three plans for the customer's choosing:

- Payment of a $5.00 monthly charge with a $40.00 minimum usage per month. Western Union recommends this plan if your monthly AT&T long-distance calls exceed $75.00.
- A no-fixed-fee plan with no monthly charge, with a $10.00 minimum usage per month. The catch: daytime rates are about $0.025 higher per minute to on-network cities.
- A plan with a $10.00 monthly charge but no increase in the standard rates.

Western Union's rate structure is based on the length of time you speak, to the nearest minute, and the distance of the call. Like its competition, Western Union establishes separate rate structures for daytime calls, evening calls, and late-night and weekend calls. Rates vary depending on the type of program you subscribe to. A rough guide:

	Per-Minute Rate Range
WEEKDAYS	
8:00 A.M. to 5:00 P.M.	$0.28 to $0.50 per minute
WEEKNIGHTS	
5:00 A.M. to 11:00 P.M.	$0.17 to $0.30 per minute
LATE NIGHT/WEEKENDS	
11:00 P.M. to 8:00 A.M. and 5:00 P.M. Friday to 8:00 A.M. Monday	$0.13 to $0.20 per minute

Unless you're a big caller you can forget about discounts. If you make $750.00 in calls per month, you receive a five-dollar credit for every fifty dollars in calls. The credits, however, can only be applied against a limited number of Western Union services.

ALLNET (800–982–8888) is another long-lines company that is growing on a national level. In some areas its rates are competitive with MCI's and Sprint's, but not yet as good as SBS Skyline's. It doesn't hurt, however, to see if it serves your area and compare its rate structure with the others.

The Davids Also Do Battle

In addition to the companies that have national hookups, there are also a host of local long-distance companies. These smaller operations lease AT&T's lines on a long-term discount basis. They can afford to re-rent these national lines to their local customers at a slight markup. Everybody profits as long as the smaller companies continue to provide a service that is either cheaper or more efficient than the major national long-lines companies. With competition on the increase it is but a matter of time before many of these smaller companies drop by the wayside. In the meantime, however, they are certainly worth checking out. If you live in an area served by one of these companies, give them a call and ask for their literature.

Some of the larger independent long-lines companies that serve a specific region:

SELECTED STATES:

Alabama
Interstate Communications
404–645–1013

Florida
Microtel, Inc.
305–422–9932

SELECTED STATES (*cont.*):

Georgia
Interstate Communications
404–645–1013

Kansas
Business Telephone Systems
512–456–1100

Louisiana
Business Telephone Systems
512–456–1100

Ohio
Mid Atlantic Communications Corp.
800–682–8282

Oregon
Savenet
503–241–0090

Tennessee (portions of the state)
Interstate Communications
404–645–1013

Texas
Business Telephone Systems
512–456–1100

Wyoming
Wylon, Inc.
307–266–3627

Southwestern States
U.S. Telephone Communications, Inc.
214–741–1957

SELECTED CITIES:

Chicago, IL
Allnet, Inc.
312–443–1444

Dayton, OH
Dial America
513–224–9764

Knoxville, TN
Call U.S.
615–525–2666

Miami, FL area
Teltec, Inc.
305–932–3031

Milwaukee, WI area
Telecommunications Services, Co.
414–784–6752

Pittsburgh, PA
Penn Telephone Inc.
412–443–9500

Rochester, NY
Alternative Communications Corp.
716–442–2904

Seattle, WA
Altcom, Inc.
203–328–0567

St. Louis, MO
LDX, Inc.
314–621–1199

Washington, DC area
TDX Systems, Inc.
703–790–5300

SHOPPING FOR A LONG-DISTANCE SERVICE

With the split-up of AT&T there is a great deal of pressure on all telephone users to make a decision concerning their long-lines service company. There are a few items to consider and a few ways to cut and control your telephone costs.

First, if you make the vast majority of your calls to friends and relatives in your own town, there isn't much need to buy a special long-distance service. Remember that some services charge you a fee even if you never pick up the phone. SBS Skyline doesn't charge a monthly fee but does impose a fifteen-dollar-per-month minimum on your long-distance calls. If you elect not to go with one of the many competing long-lines companies, you will probably be assigned to AT&T. Not a bad decision. After all, at present, Ma Bell doesn't impose a long-distance charge unless you actually make a long-distance call. There is talk, however, of local access companies imposing a flat rate on all customers to be hooked into the AT&T long-lines network.

If you do start shopping for a long-lines service—and it is important to shop and compare for yourself—here are a dozen questions to ask when evaluating your home communications needs. They may save you quite a bit of money.

1. *Are there any hidden costs?* If you have to pay message-unit charges for access to your long-distance service, this adds to the cost of your call. And if you don't ask your long-distance service company if it charges for message units, you may be billed for these charges without knowing it. Some long-distance services, SBS Skyline for example, will even refund your coin from a pay phone.

2. *Is there a simple, understandable rate structure?* Why are companies so leery about telling you what it costs to use their services? When you ask a prospective long-

lines service company to send you a brochure explaining its operation, does it spell out in detail exactly what its charges are? The chances are that a few will, but others tell you everything but what it costs. You will be shown a beautifully illustrated chart comparing its service with the competition. But you don't get a clear, concise price list. Some companies have taken a page from the airlines' book. Ask two people from the same long-lines service company what it costs to call from your town to Aunt Tilly in Davenport, Iowa, and you get two different answers.

3. *How are you charged for fractional minutes?* Many long-lines companies, including AT&T, round up to the next minute. This means that if your call is only a few seconds past one minute, you're charged for two full minutes. Some services bill only to the next tenth of a minute, meaning that you can save the cost of fifty-four seconds on a call that is six seconds or less over the minute mark.

4. *Do you keep a record of often-dialed long-distance numbers?* If you don't, it is a good idea to start. Now that Ma Bell is stripped of her local telephone businesses, she is making the most of her opportunities. In the future, you will be billed for long-distance information calls. After all, why should Ma Bell give away a service for which she can just as easily charge? Too many people were using AT&T's lines to obtain a long-distance number, only then to use Sprint or MCI to make the actual call. Call it Ma Bell's revenge!

5. *Do you have all records or documents in front of you before you make a long-distance call?* Remember, time is money. And if you don't buy this old maxim, wait until you receive your next phone bill. Know what you're going to discuss before starting to dial. Also, have an alternate party you can speak with so the call won't be wasted if the person you want is out.

6. *Do you have to make an operator-assisted call?* If you are using AT&T's lines, you will be billed an additional

three-dollar surcharge for each collect or person-to-person call. Sometimes it's better to gamble that your party is in and pay just a few cents for a station-to-station call rather than that heavy surcharge.

7. *Do you use a credit card or make third-party-charge calls?* Aside from the loss or abuse factor, credit cards, though convenient, cost money. With every credit card call or charge to a third party, AT&T hits you with a surcharge. Many businesses have dropped their phone credit cards and have their employees include phone calls on their monthly expense accounts.

8. *Do you pay a monthly fee to use your long-distance service?* Some services charge you a monthly fee to use their service. Some, like Sprint and MCI, have dropped their monthly charges; SBS Skyline doesn't charge a monthly fee but does impose a monthly minimum of fifteen dollars; AT&T has no separate charge or minimum fee.

9. *Do you make your long-distance calls when the rates favor you?* There is a big cost difference between a call made at 4:55 P.M. and one made at 5:01 P.M. Why not plan your calls to take advantage of the most favorable rates? Also, some long-distance companies base their rates on different time periods. Find the time breakdown that suits your needs, not the company's.

10. *Are you being charged for equipment you don't use or need?* Unless you own your own telephone you are being charged a monthly rental fee for each and every piece of equipment in your house. Of course, you can keep as much or as little of your local phone company's equipment as you'd like. But what you turn back to the company you won't pay for. You be the one to make the decision. Take inventory of your needs and decide if the present equipment meets or exceeds those needs. In some homes there may be more than one trunk line. Do you really need more than one incoming trunk line?

11. *Do you know your local rate structure?* The front

pages of your telephone directory explain your local rate structure. Compare your bill with these rates. Are you being fairly charged? If you have any questions about your rates, call your local telephone company's business office. If you find discrepancies, ask for a refund or adjustment.

12. *Do you use your telephone for business?* If you can show that your home telephone is used to generate any part of your income, you may be able to take a tax deduction not only on your long-distance and local service but also on equipment rental. If you bought your own equipment, you may be able to depreciate a portion of it. Contact your accountant or local IRS office if you have any questions about deducting your telephone service. It can be a major savings come tax time.

IT PAYS TO COMPARE

How can you compare long-distance services? Cost, obviously, is a critical factor. However, unlike shopping at the local supermarket, it isn't always easy to find out what your actual costs are. Call a long-distance service and you often get the runaround. Sure, they are anxious and eager to sign you up as a customer, but not always eager to give you specific answers when you dig into their charges. Look at the charts showing the various possible charges and services offered by the major long-lines companies. They may give you some ideas as to questions you can ask when you start shopping. Remember, during the competitive fight to sign you on as a customer, these features may vary from region to region and from time to time. Companies often drop their initial setup charges when starting a community marketing campaign.

As for the cost of actual calls, keep in mind that in some cities a message-unit charge is imposed by your local phone company for reaching the access number to a long-lines computer. Neither AT&T nor SBS Skyline impose these message-unit charges. A sampling of daytime (8:00 A.M. to 5:00 P.M.) calls to various parts of the country from New York City looks like this:

To	*Minutes*	*AT&T*	*MCI*	*Sprint*	*SBS*
Big Sky, MT	4.1	$2.40	$2.23	$2.14	$1.67
Charleston, SC	4.4	2.34	1.95	2.09	1.67
Chicago, IL	4.2	2.34	1.95	2.09	1.67
Dallas, TX	5.5	2.84	2.39	2.53	2.04
Hartford, CT	6.2	2.79	2.40	2.51	1.76
Los Angeles, CA	4.8	2.70	2.24	2.42	2.15
Miami, FL	6.1	3.28	2.79	2.92	2.42
Wausau, WI	2.4	1.48	1.34	1.34	.93

Aside from the long-lines services listed in this book, don't forget to check the Yellow Pages issued by your local telephone company for local independent services that may be operating in your area. To obtain the number of national companies, first check to see if there is a local 800 number available. Why should you pay for the call? They are beating the bushes looking for your business. For 800 information, dial 1–800–555–1212.

Request all the company's literature and read it over. It's a good idea to have a list of questions in your mind when making a comparison. Finally, call a company representative and ask any questions the brochures don't answer for you. If your neighbors or friends are using a service, ask them how they like it. There have been some complaints that reception isn't as good with some companies as with others.

In testing all the major services I have found that the reception, overall, has been satisfactory. Problems occur not so much in communicating as in reaching your party. With all the non-AT&T services, I have gone through the multi-dial process only to reach an intercept recording informing me that I was not getting through. Reaching any recording is annoying. It was particularly so when I just raced through a twenty-digits-plus dialing exercise. The lack of a warm human voice in time of need is also a frustrating factor that the long-distance services lack. Ma Bell may have been a tight-fisted old dame, but at least she was always there when you needed her. Another problem that has irked me is not being able to get through to the initial access number. A busy signal or, worse yet, dead silence, is often enough to edge the frustration factor close to the breaking point. Again, Ma Bell has an edge.

Obviously, cost is the prime factor that leads one to leave Ma Bell to her rocking chair in quest of cheaper services. If cost were no object, AT&T would still provide the best overall service. However, since cost is a key reason for buying a long-distance service, I have found that SBS Skyline offers the best consumer service for the dollar. But for the business person who is often on the road and using hotel and motel phones, the added cost of Skyline's four-dollar charge for each network city may be enough to swing the balance to Sprint or MCI. Fortunately, you do not have to make a long-term commitment.

A few general observations:

✆ Regardless of which service you choose, a general rule is, the longer you talk, the less you pay per minute. If you're a real phone gabber, you'll save more using the long-lines services compared to AT&T. For short calls, AT&T's high first-minute rates make it an easy target for the competition. Compared with Ma Bell, your biggest savings with the long-lines companies is during the evening and at night.

✆ The greater the distance, the greater the savings. Exception: SBS Skyline charges the same for calls to all non-neighboring states.

✓ For short distances of less than a hundred miles, the long-lines companies' rates may actually be *more* than AT&T's. Check and compare with your local phone company.

Shop around. Decide for yourself which service best suits your needs and those of your family and give one a try. If you're not happy, try another.

Good luck!

—4—

Planning Your Complete Home Communications System

Since the AT&T split-up the manufacturers of telephones and long-distance dialing services haven't been the only ones interested in harvesting the bonanza of consumer dollars flooding into the home communications market. Suddenly, consumer magazines, daily newspapers, and the business press are dotted with advertisements encouraging homeowners and apartment dwellers to rush out and buy everything from cordless phones, answering machines, and auto-dialers, to a cellular phone for the family car. Aside from providing the nice profit that the telecommunications manufacturers stand to gain, installing new communications equipment in your home may actually save you time, money, and a great deal of aggravation.

Your equipment can include any of the following:

✆ *Home Telephones.* You have the widest possible selection at your disposal. Your choice is practically unlimited. Rotary dial or true tone? Wall or desk, small or large, cordless or wired? The styles and varieties of telephones provide you many options. Some phones come in decorator shapes and colors, while others look like throwbacks to the early phones that your grandmother first used.

✆ *Mute or hold button.* This feature allows you to speak to another person in the room without being overheard by your caller.

✆ *Redial button.* Reach a busy signal? Hang up and wait a few minutes. Instead of redialing the same maze of numbers, just depress the redial button.

✆ *Memory.* This feature lets you preprogram one or more numbers into your telephone. A doctor? the police? the office? Store their numbers in the memory bank, and you only have to press a single button instead of up to twenty-five.

✆ *Ringer.* Is your bell too loud or soft? Adjust it with another button.

✆ *Answering machine.* You no longer have to miss those important calls when you're away from home or when you're just not interested in answering that infernal ring. An answering machine can be a separate unit or part of a combined telephone-answering-machine unit. You can retrieve messages when you return home or, if you're away, just call your own number and the answering machine will relay your messages to you. You used to get your away-from-home messages only if you carried a separate cigarette-pack-sized box that, when placed next to a telephone mouthpiece, sent an electronic signal to activate the answering machine and play back the recorded messages. No more! Now all you have to do is call your home and, using a true tone phone, punch in a few preselected coded numbers. Within a matter of seconds not only do you get your recorded messages, but a sexy voice also tells you the time and date the message was received. Forget a message? Hit another coded number, and the answering machine returns to the last message and plays it again.

✆ *Auto-dialer.* If you make long-distance calls and subscribe to one of the long-lines service companies, an automatic telephone dialer is a great convenience. Rather than punch in twenty-five separate numbers and hope you don't make an error along the way, just punch one, or at most two, buttons on your auto-dialer, and your electronic signal is sent on its way. Programming an auto-dialer is a simple procedure. Some phones now come equipped with auto-dialers that can hold several dozen numbers. Changing a number is as simple as changing a station on a memory radio. It's all done with the miracle of electronic switching. If your needs call for an auto-dialer that can hold as many as a hundred separate numbers, no problem. A device about the size of this book is all that is required. Connecting one of these devices to your telephone is also no problem, as long as your phone is equipped with a new modular plug. More about these later.

✆ *Intercom.* You're in the basement and the call is for that teenager upstairs. Instead of yelling up the stairs and trying to compete with the sounds of the Rolling Stones, a push of an intercom button on your telephone extension rings a buzzer in the teenager's bedroom. Of course there is no guarantee that the buzzer's sound will overcome the sound of music, but it's still a better bet than wearing out your lungs screaming in the basement. Intercoms are not as common on corded phones, but they are easily accessible on the cordless models. It is one of their many advantages.

✆ *Pager.* Pagers are becoming the latest rage. Although not truly a telephone device, they still fit into the broad category of electronic communications equipment. The problem with any pager is that you need to work through a third party. Basically, a pager is a small radio receiver. When you want to send a message you simply contact a transmitting station, give someone your message, and the station sends a unique signal to the paging device. You can buy a receiver but still must make an arrangement with one of the paging companies to serve as your transmitter. This usually requires a small monthly service charge plus a fee every time you send a message. Initially, pagers were designed to operate within a fairly small area—your city or town. Now there are companies that rent small pagers that will pick up messages bounced off satellites anywhere in the world.

✆ *Speaker phone.* How many times have you had your hands full and just didn't need a telephone connected to your ear? With an amplifier, or speaker phone, you depress a button and do all your talking from afar. The calling party's voice is amplified and a very sensitive microphone in the telephone picks up your voice or any other voices in the room. You carry on a normal conversation. Well, almost normal. Some speaker phones still sound as if your voice is originating in a tunnel, but at least you can get your message across without ever having to hold the phone. If you don't mind tying up your line, it is possible to combine the intercom and the amplifier features. Thus, a phone can be placed in a child's room with an open line in your bedroom. Any cry or sound is received.

✆ *Fire, smoke, and health monitor.* Hook a smoke/fire detector directly to your telephone lines. If there is a problem, the system can be programmed to contact you if you're at another location, as well as to place a direct call to the local fire department. For the elderly living alone a small box, about the size of a toothpaste case, is carried in a pocket or purse. In case of any medical emergency just press a button on the box and a signal is transmitted through your special telephone attachment. An emergency call can be sent directly to a hospital or to any party whose number is preset in the transmitter.

✆ *Computer hookup.* With the rapid growth of home computers, the telephone lines are an important link to other computers, as well as a lode of information mined from any of the many commercial and private on-line data bases. In essence, your home computer is plugged into your telephone system by means of an electronic device called a "*modem.*"

PLANNING YOUR COMMUNICATIONS SYSTEM

Before running out and buying one of the above components, take a moment to think about your total communications picture. If you're going to buy a new phone, have you checked to see if your present home or apartment phone plugs are compatible with the new modular plugs that are on all new telephones? Will your phone do everything you want it to do? For that matter, what do you want your phone to do for you besides those old-fashioned things like calling a friend? Is it legal for you to install your own wiring, or must your local phone company be called in? In most states it is now perfectly okay for you to do the work without paying a local operating company to do it. However, laws do change. Be safe and give your local company a call to be sure.

Are there existing telephone wires or outlets where you want your new equipment to be, or will you have to install them? If you're building a new home or adding an addition, seriously consider including a modular outlet in every room. You don't have to use it now, but who knows what your communications needs will be in a few years. Besides, the cost of wiring during any construction period is relatively inexpensive. Think what a few dollars spent now will do to your home's resale value.

SHOPPING FOR PHONE EQUIPMENT

A few tips on shopping for your phone equipment:

1. Don't assume that you have to buy the phones already in your home. Remember, it is now your choice. You have the right to keep them on a rental basis, as before the AT&T breakup; buy them outright from the local phone company; or turn them back to the phone company and

replace them with any equipment you decide you want.

2. Read the warranty. They all differ. Some manufacturers give a ninety-day warranty, while others last for a year or more. Does the warranty cover both labor and parts? If you are going to buy a refurbished phone, does it carry the same warranty as a new phone? Is the warranty issued by the manufacturer or the dealer? If it's a dealer warranty, who will fix it if it breaks? Most dealers know little if anything about telephone repair. Beware! For that matter, it is important to find out who is responsible for repairs. Will you have to ship your phone across the country and wait six weeks, or is there an authorized repair service in your area? If it's your only phone, will the repair service give you a "loaner" while your link to the world is in the shop?

3. Try out your new phone before buying it. All new equipment comes with a modular plug. Even in a dealer's shop it should be possible to plug a new phone into an outlet. A major phone center will have several phones already hooked up. You're used to Old Reliable at home. Don't assume that all new phones will sound alike. Is the sound clear? Can the person at the other end hear you clearly? If you're about to buy a cordless phone, see how far away from the base unit you have to be for the phone to operate. How easy is it to use the buttons or dial? Do they stick?

4. Does the phone have an FCC number? If not, you may have a problem. (See page 54.)

5. Is the equipment well made? Obviously, you can't tell about the guts of the system by looking, but looking will tell you how well built the cover and exterior are. For example, is the bottom smooth or completely unfinished? A jagged edge will scratch your furniture. Is the cord wound up like a spring, or does it release smoothly? With a tightly coiled cord you may wind up with your phone embedded in your ear every time you make a call.

6. If you intend to buy a phone with tone dialing, be

sure the equipment you select gives you true tones. All push-button phones are not alike. Some give you a true tone that is compatible with long-distance services like Sprint, MCI, and SBS Skyline; others operate on a rotary-type system within the phone but are not compatible with the long-lines companies.

7. Does the equipment have a modular plug? And are your home phone lines wired to accept the new modular plugs? If you have the older four-pronged outlets, you can use all the new equipment but you will need an adapter. (See pages 66–68.)

8. Shop around. Compare prices and equipment. Phone equipment is not price fixed. You can get bargains.

— 5 —

Wiring Your Own Telephone Equipment

Before getting into the nitty-gritty of wiring, a few words about terminology are in order. Don't be afraid of or confused by technical words. They are fairly simple and straightforward. Here's a very brief glossary of the more common terms you'll come across:

Outlets: Connectors, either in the wall or along your baseboard, into which phone lines are plugged. An outlet can have a wall plate attached to it like your light switch or it can look like a matchbox on the baseboard with four small holes in it. In the past, all extension phones had a four-pronged plug attached to the wire stretching from the phone to the wall. The four-pronged plug was inserted into a matching outlet containing four corresponding holes. All telephones and telephone equipment today use what is called a "modular format." Modular outlets contain a single square-shaped hole into which a matching plug is inserted. Not only are they easier to use, but supplementary equipment such as answering machines, auto-dialers, and cordless-phone receiving boxes are all compatible.

Jacks: The interfaces between a pair of connections. The

Surface-Mounted Jack

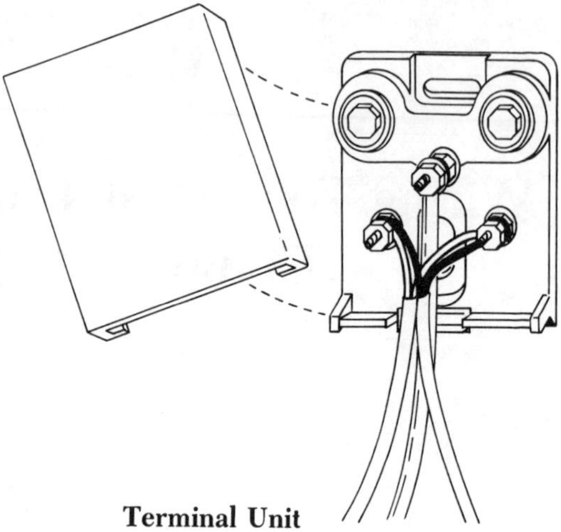

Terminal Unit

modular jack, for example, is a wonderful invention that eliminates the problem of which wire goes on what terminal—it's all done for you, ready for you to insert into a modular plug.

Standard Network Interface (SNI), often called a Network Interface (NI) or a Demarcation Point Annex (DPA): The point at which the telephone company's wiring stops and yours begins. The SNI is usually located at some point inside your home and usually contains a modular jack. If you don't know where your SNI is, look at the outside of your house and see where the phone lines enter. At an interior corresponding point, you should find the SNI. In most homes the SNI is located on an inside basement wall near your power meter box.

Wires: Generally telephone wire is 22–24 AWG in size and usually contains four color-coded wire strands: green, red, yellow, and black. Each colored wire, or conductor, is for a purpose. The red conductor makes your phone ring; the green provides operating current; the yellow carries a low-voltage current to illuminate a dial light; and the black serves as a ground. The application of these colors is not a hard and fast rule. The important point to remember is that, when connecting one wire to another, keep the colors consistent. In other words, don't attach a green wire to a plug with a red marking. Also remember that telephone wires do carry a small electrical voltage. If you plan any major wiring or rewiring, first disconnect the inside wire at the SNI. There is another valuable safety hint: Do take a phone off the hook when you're working on an adjacent wire. Removing the phone cuts back on the small amount of voltage that is passing through the line.

If you are planning to install your own wiring, it's a good idea to keep these *don'ts* in mind:

Don't ever use telephone wires to carry anything but telephone signals. Your standard 110-volt home electrical system must never be interchanged with your phone system.

Don't work on telephone wiring during a thunder and lightning storm.

Don't work on telephone wiring if you wear a pacemaker.

Don't connect more than five telephones to any one line.

Don't string telephone wire through any damp or potentially wet areas in your home. If your basement tends to leak, keep wires off the floor.

Don't splice one telephone wire to another. To connect two wires use a jack or adapter.

Don't string telephone wires through the same conduits containing other types of electrical wiring.

Don't string telephone wires next to hot-water pipes, power lines, transformers, radio antennae, or outside the home. The chart below may provide a few guidelines:

Electrical Source	*Description*	*Minimum Separation*
ELECTRICAL WIRES	Bare light or power wire.	5 feet
	Open wiring not over 300 volts.	2 inches
	Wires in a conduit or in nonmetallic sheath cable, or power ground wires.	none
RADIO OR TELEVISION	Antenna lead or ground wires	4 inches
COMMUNICATION WIRING	Community television systems; coaxial cables	none
SIGNS	Neon	6 inches
LIGHTNING SYSTEMS	Lightning rods and wires	6 feet

DATA: GTE Communication Systems Corporation

Here are a few *do*'s:

Do comply with all building and electrical codes in your area.

Do unplug the inside wiring from the SNI when doing any major wiring or rewiring.

Do unplug or take a phone off the hook when you're working on it or its wiring.

Do use caution when drilling through walls or ceilings. You don't want to drill through existing electrical wires or cables.

As for the planning of the actual wiring, it's a good idea to make a floor plan of your home or apartment before you start. With your needs firmly in mind, decide where in each room you want your communications equipment to be located. Then, starting from either an existing outlet or the SNI, draw a schematic plan hooking your planned system together. If two lines cross each other, you need a *bridging adapter* to interconnect the lines and prevent a spaghetti-like maze of wires crisscrossing your home.

Since in some instances you will have to snake wires through walls and ceilings, decide well in advance the easiest route to follow. Also decide in advance which type of outlet is best for each location. If you want a kitchen phone wall mounted, for example, you will need to install a wall set connecting block. This unit is like the electrical wall boxes except that it carries phone, not electrical, lines. It also has a simple modular connection.

Select your tools. A basic complement includes:

Wire cutters.

Two screwdrivers with insulated handles: a small screwdriver for those tiny screws and a larger screwdriver for larger screws.

Wire strippers.

Drill and appropriate bits.

Saws: you'll probably need a keyhole saw if you're planning to transit walls.

Snaking equipment for carrying wires through walls and ceilings.

Before you start planning to include a new phone in every room, keep in mind that the phone company only supplies a certain amount of power through a single line. The power is enough to operate about five phones. The key to determining how many telephones will work in your home is the ringer equivalence number, or REN, of each phone. This number can be found on the registration label of all phone equipment. It is usually printed in small type, naturally, on the registration label stuck to the bottom of the equipment. The older phones require more electric current and have a REN of 1 or more; the newer models often have a REN of less than 1. Some cordless phones, for example, have a REN of less than 0.5. If you have any questions about the power input or the number of RENs available to you, call your local telephone company office.

If, for example, you have five older model phones, each with a REN of 1 or more, you'll either have to make some compromises in the selection of your new additional equipment, replace the older equipment with phones with a lower REN, or consider bringing in another telephone line. Obviously, a second line means twice the monthly line rental.

There is another solution. If you have two phones close to each other, there isn't much purpose to having them both ring. By disconnecting the wire that causes one of the two phones to ring, you also cut down on the voltage necessary to operate that phone. If you're dealing with an old connection, just open the jack and unhook the *red* wire. Now your phone won't ring and you have solved the REN problem. If you want to disconnect the ringer on your new phone, you either have to unhook the *red* connecting wire in the phone itself or in the lead connection going to the modular junction box on your wall or along the baseboard.

In addition, you may note a letter following the REN number. This letter designates the frequency of the ringer within your phone. Most telephone companies supply Type A ringing current, which is 20 and 30 Hz. If you see an "A" following your REN number, you'll be perfectly safe operating your phone on any single-party line. If you discover another letter, call your phone company before buying your new equipment. The chances are you'll have no problem.

You should also look for a statement next to the REN number that will let you know if your present equipment or the phones you want to buy are FCC registered. It is important to assure that your equipment is approved by the Federal Communications Commission because of the standards of quality established by this agency. If you don't have equipment carrying the FCC stamp, you may have a few problems. The obvious problem is the fact that your equipment may not work. It may also affect all the other phones in your home, putting them all out of commission as well. Further, it is against the law to connect any phone or telephone equipment to your phone lines that is not approved by the FCC.

If you can't find any mention of the FCC on your old equipment, don't assume it hasn't been approved. If it's an older piece of equipment in your home, you know it works. But since you want to stay on the right side of the law, have your older phones checked out and certified as FCC approved. You'll have to get in touch with a repair service or phone refurbisher authorized by the FCC to approve your equipment or make the necessary internal adjustments to meet all federal standards. Once these standards are met the authorized agent will place an "FCC Approved" registration label on your equipment, and you're in business. To find a service company that can undertake the job, give your local FCC office a call. Their number is in the phone book in the section listing government offices. If you live in an area that does not have ready access to a local FCC district office call the commission in Washington (202–632–7000 or 202–632–7260), and they will give you the names and addresses of the authorized businesses closest to you. You can have your phone altered or refurbished by many services, but unless they are FCC approved you won't receive certification that your equipment is legally acceptable by the government. FCC approval doesn't imply quality. It simply means that the approved hardware meets FCC technical standards.

If you have been using AT&T's equipment for years, it obviously meets FCC standards even though it may not carry the "FCC Approved" label. Equipment installed by the phone company many years ago or equipment you may have installed with the phone company's knowledge and approval before July 1, 1979, is perfectly legal as long as you either keep paying your rental bills or buy the phone outright from your local phone company. This older equipment will appear on an FCC "Grandfathered Equipment List." Again, you can check on the status of your equipment by contacting the FCC. They maintain a list of all grandfathered equipment. It's available by sending five dollars to the U.S. Superintendent of Documents in Washington, D.C. However, rather than buying a five-volume list of numbers, it is much easier, not to mention cheaper, to call or write the FCC and ask whether your specific telephone equipment meets their standards. (Write: Federal Communications Commission, 1919 M Street, N.W., Washington, D.C. 20554. The FCC also maintains a special number for questions about consumer-owned equipment: 202–634–1833.)

Failure to list your equipment with the local phone company won't mean a visit by a friendly FBI agent; you won't go to jail or become a face on the post office wall. The odds are the phone company will never discover your subversive equipment. However, if they ever decide to check their customers to determine whose equipment is listed and whose isn't, they don't have to make house calls. Electronic sensing equipment hooked up to their switchboard will spot your unlisted equipment in no time. What you may think is a mistaken ring may really be a phone technician measuring RENs in your equipment and comparing electronic data against company records. New electronic searching methods can even locate unlisted equipment in your home without your phone's ringing or your ever knowing you have been investigated.

If you are caught with unregistered telephone equipment in your home, you will probably be asked to comply with local and federal regulations. That should be the end of the matter. If, however, you want to play games with the phone company, remember that they have the last move. Either play by their rules or hook your phones to someone else's lines. Ma Bell may have been broken up, but wherever you live there is still one—and only one—phone company.

The monopoly hasn't been destroyed, just cloned.

MAKING THE CONNECTION

If you live in an older house or apartment, the chances are you have a network of telephone wires snaking about your home. Don't discard them or figure that because they are old or haven't been used in years they won't work. Technology has made major advances, but those old telephone wires will do just fine no matter where they are hidden or how many times they have been repainted. Of course they won't have simple modular connections, but these can be added quite easily.

Always use wiring already in place. It has two things going for it: It was installed by a pro, and it works. However, don't let the thought of either rewiring your home telephone system or connecting new equipment to old wires scare you. It really isn't very difficult. And when you're finished you will not only have saved quite a bit of money but you will gain tremendous satisfaction from doing what you may have thought was the impossible.

Using Existing Wires

First, check every inch of your old wiring for breaks. Many times an old wire link has been cut or removed entirely from a room. It would be a shame to go through the frustration of hooking up your entire house with new communications equipment only to discover that it didn't work because of a severed line.

You may also discover a terminal or junction box hidden in the recesses of a closet or buried on the baseboard behind the couch. There is a good chance that one of your present telephones may be wired to one of these terminals. If so, fine. You're ahead of the game already.

If you are going to connect a new piece of telephone equipment (phone, answering machine, auto-dialer, etc.) to that old piece of wire or junction box, you first have to get at the wires. The small junction box cover will either snap right off or can be unscrewed. Obviously, if you see a small screw in the dead center of the box, don't try to unsnap it. A few twists with your screwdriver will unseat the screw. If the cover doesn't pull right off when unscrewed, the odds are that years of paint have fixed it to the wall. No worry. Slip the screwdriver between the lid and the wall and you should be able to snap away a few decades of paint. With the interior of the box exposed you will discover that your phone is connected to the terminal by two, three, or four screws. Most phones have two or three connections.

Unscrew the old phone you are going to replace. Then, since you want to replace the old terminal with a new modular terminal, remove the old terminal with a pair of wire cutters. Be sure you leave yourself enough wiring to add the new terminal—about three or four inches. If your old instrument is in working order, you know the connections are good, as is the wiring. However, it is a good idea to remove from the exposed wires or connections the layers of dirt or oxides that may have accumulated through the years. A piece of fine sandpaper or emery will do the trick. An old emery nail file can be used if you don't want to run out and buy sandpaper. Rub the wires and connections until they begin to shine. They are copper and will have built up a thin layer of green copper oxide.

At the base of your new modular outlet you will find four screws that correspond to the screws on the old terminal box. You will also note on the back of the new connection either the letters R, Y, G, and B or their corresponding colors, red, yellow, green, and black. Now, simply screw the color-coded wires to their matching con-

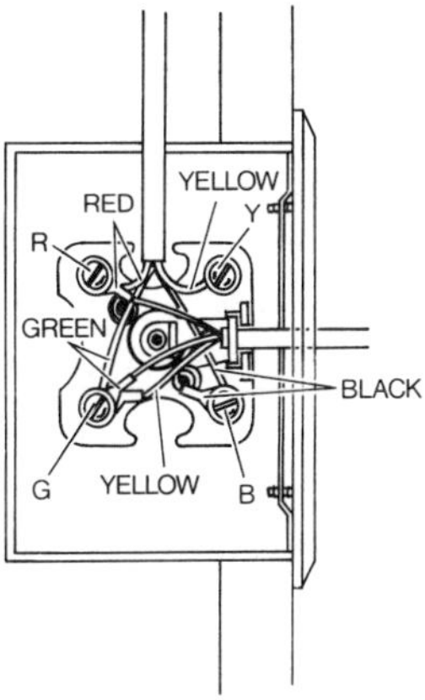

Connections for Desk Telephones

nections. If you have any doubts about which color goes where, draw yourself a diagram of the wiring hookup from the old terminal and match the identical setup on the modular terminal. It really isn't hard.

Finally, screw the modular terminal onto the baseboard, or wherever you want to replace the old terminal. You are now ready to plug in your first piece of telephone equipment. Now that wasn't so bad, was it?

Adding New Wiring

In many homes, both new and old, there is a need to add new wiring. Sometimes you may have to add a new section of wiring to existing lines or start from scratch. If you are in a home without any telephone wires, which is a rarity, you will have to start from the point where the local phone company's lines leave off—the SNI. However, whether you start from the SNI or want to add a length of new wire to Ma Bell's old existing cable, the principle is the same.

Before you start wiring your home to accommodate a slew of new phones, keep in mind several factors. First, the number of phones you can add to an existing system will depend on the ringer equivalency numbers or RENs of all the phones on your line. Exceed 5 RENs, and you may have a problem. (Disconnect a *red* phone wire at the jack or the phone itself and you can cut back on those RENs; you will also eliminate the phone's ring.) Second, the longer the length of wire running from the SNI, the more resistance you add to your total system. Chances are you won't be adding that much more new wiring to make much of a difference, but it is a consideration.

When adding new wire, don't try to salvage old telephone wire that is not being used. Twenty-four-gauge, four-conductor telephone wire isn't all that expensive. Buy the best and stick with it. Start wiring from the SNI. *But don't connect your wire to the SNI before you start to string it through your house.* Hooking into the phone company's main lines is the last step. Besides, even though the current is very slight, why handle live wire if you don't have to? Thread your new wire along the most direct route to the room in which you have selected to place your new equipment that is farthest from the SNI. Don't worry about other rooms along the route. You can add these later by running a separate wire from the primary line. You will probably have to thread your wire through floors, walls, and ceilings. Select the route well in advance. Keep your wires hidden and, if possible, out of sight. They are thin enough to fit easily under most baseboards, behind moldings, over door and window sills, and, when necessary, under wall-to-wall carpets.

When you have your wire loosely strung from the SNI to the farthest point, the time has come to test your work. If you're planning to add modular equipment, attach a modular connector onto the end of the phone wire. The connector need not be attached to a wall or baseboard. On the back of the socket you will see four screws with their color-coded identifiers: R for red wire, Y for yellow, G for green, and B for black. Attach the new wire to its appropriate connector with a screwdriver. Now is the time to connect the wires to the phone company's terminal box, or SNI.

Be sure that you are about to hook into a phone box. Look on the side of the box for a Bell, AT&T, Western Electric, or other phone company designator. Obviously you don't want to get involved with the electric utility's equipment. (Their wires must not be exposed, whereas phone wires may well be out in the open.)

If you have any doubts whatsoever about whether you are working with a phone junction box or an electric outlet, back off. Call your local utility company and ask them to show you exactly where their equipment is located. It's better to be safe and sure than sorry! When you know you are staring at a phone junction box, swing back the cover or, if the cover is held on with a screw, remove the cover with your screwdriver. Again, connect the color-coded wires to their appropriate seats. The chances are that colored wires will already be attached to the only color-coded screws at the SNI. Nothing to worry about. They are your present home phone wires. It is perfectly all right to add your new color-coded wires atop the existing wires.

With one end of the new wire now connected to the SNI and the other to a modular connector, plug in your new modular phone. Hear a dial tone? Congratulations, you did it! If you don't hear a dial tone, don't panic. Before you start ripping out your new wiring, why not test your new phone? Maybe the problem is with the equipment rather than your good wiring job. Take the phone to a friend's house and plug it in. Does it work? If it does, back to the wires. Check all your connections. Assuming your wire is in working order, and if it's new it will be okay, the problem has to be in the connections.

Now that you know your primary line is in good working order, it can be attached permanently to the floor, baseboards, etc. For years, phone people have used special insulated staples for this purpose. You can buy phone staples at any large hardware or phone supply store. *Don't* use those thin wire staples that are in your desk stapler. And don't use a high-powered carpenter's staple gun. You can shoot a staple right through the wire. You can buy a special staple gun for phone staples for about twenty dollars. The gun makes the work much easier, but a small hammer will also do the job. With the wire kept taut, place a staple every three to four feet. At corners, place a staple directly in the corner to prevent the wire from bulging. When the new wiring is firmly in place you are ready to add extension spurs from the main line.

TYING IT ALL TOGETHER

Fortunately, there are many companies that manufacture devices that allow you to perform many different functions with your existing lines. Once you have your telephone wires in place, whether they are designed for modular installation or not, you can always include the various add-ons. For example, let's look at some of the many devices you can buy to make your telephone hookup a little bit easier.

Wire Junction Box

Once you have strung your main or primary line throughout your house, you probably want to add spurs to other rooms. Never cut and splice telephone wires the way you often do with electrical lines. Rather, use a junction box. This small device, about the size of a box of cough drops, allows you to add extensions without cutting wires. It uses what is called a "split beam connection" so you don't ever have to strip the insulation from your primary telephone line, or conductor. AT&T manufactures two units that each sell for under ten dollars: Model 742A comes with a modular plug and allows you to add as many as four extensions; Model 742B will allow you to add three extensions. You can also pick up a unit similar to the AT&T lesser-priced model at your local Radio Shack for $4.95.

Modular Jack Converter

If you are adding new modular equipment to your present home or apartment, or if you simply want to add more wiring to an existing nonmodular system, you will need to use a modular jack converter. First, look around your house and you will probably see a small boxlike device about two inches square placed along your exposed telephone wire network. It will probably be held to the baseboard or wall by a single screw. Unscrew the cap and you will see that two, three or four colored wires are held to a metal plate by another screw. Remove the screws and the terminals' base unit. Now all you have to do is attach the color-coded wires to their color-coded terminals on a new modular jack converter. These new converters have easy-to-use, color-coded snap-ons for an extremely simple conversion. When you reattach the new converter to the wall or baseboard you're all ready for modular equipment. Modular jack converters can be bought for less than four dollars at any phone or hardware store. If you want to stick with AT&T equipment, there are three models: the flush-mounted model 725F fits against the wall; the 725C and the 725T (which has a protective swinger cover and is slightly larger than

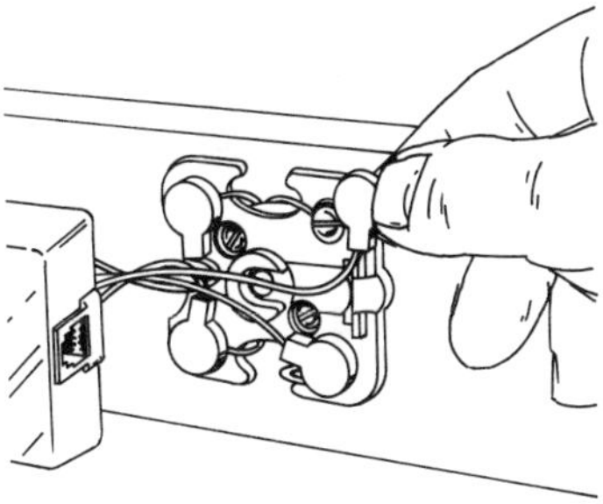

Snapping on Connectors

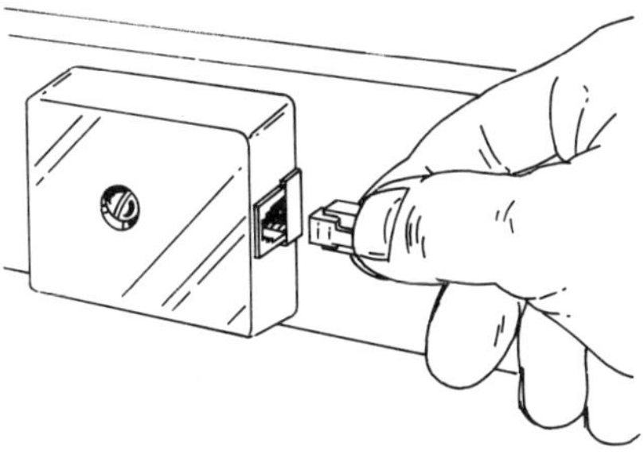

Connecting Plug and Jack

the other two models) are surface-mounted. The only tools needed to install any of these jacks are a screwdriver and a pair of scissors. Radio Shack also carries these jacks for $2.19 to $6.95.

Wall Phone Modular Jack

If you have wall phones in your house and you want to change them to newer models, you are going to need a modular jack that is designed to handle your new phones. Replacing your existing wall jack with a modular unit is a matter of unscrewing the old unit, removing it from the wall, clipping the new unit in place, and either replacing the old wall plate or adding a new one. AT&T, Radio Shack, and GTE all make modular wall plate units. AT&T and Radio Shack have white wall plates; GTE's is almond. However, the color really doesn't make much difference since the wall plate will be covered by the new telephone.

If you want to install a new wall phone, you have several choices. If you're up on your carpentry, you can carve out a small niche in your wall—similar in size to an electrical wall box, about 3½ by 2¼ inches and 1½ inches deep—then snake a phone wire from the nearest existing live

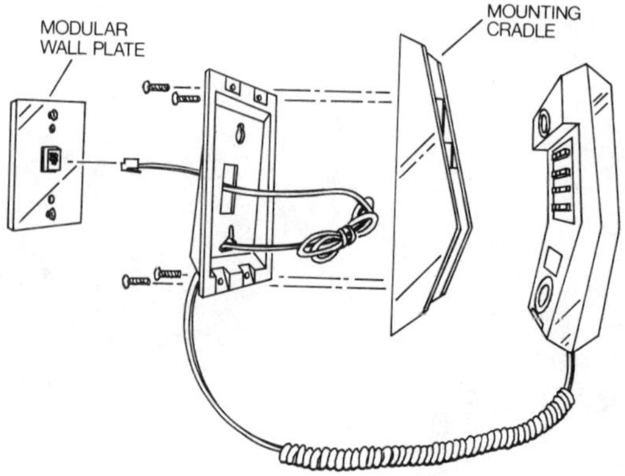

Exploded View to Show Cord Connection and Storage

phone line to the wall slot. With a new phone in place at the wall location you desire, you then attach it to a wall mount, and you're in business. However, if you're not into wall carvings, AT&T has a simple solution: Buy its wall phone mount, Model 730A; it costs less than ten dollars. No special wall insert is necessary; the mount comes with a six-foot cord that allows you to screw the mount onto any wall and insert the modular jack at the end of the cord into any modular plug along the floor. Then attach your new modular wall phone by inserting the phone's modular plug into mount, affix the phone to the mount, and you have a new wall phone. The only problem is the thin cord that runs down your wall from the phone to the floor. If you don't mind the appearance of the white cord, it is a very simple solution.

Extension Cords

If you want a new phone in a part of a room not currently served by a convenient outlet, don't think about a major rewiring job. There are such a variety of telephone extension cords on the market, you couldn't possibly have a problem one of them can't solve. Do you have a new modular phone by an older four-hole wall jack? No problem! Just buy an extension cord with a modular insert on one end and a four-prong jack on the other. Want to extend a modular phone? Your local hardware store, AT&T, or Radio Shack will sell you extension cords ranging from a few feet to thirty feet. A twenty-five-foot modular-converter will cost you less than six dollars. Want to extend the cord running from your new phone to the telephone line? Buy a handset cord. They come in a variety of colors to match your phone, are coiled, and are modular at both ends. AT&T has color-coordinated cords to match most of their phones in many lengths at under fifteen dollars. If you don't like their heavy-duty vinyl, you can also buy fabric-covered cords.

Plugs

The chances are you are going to need some type of adapter plug. Lets look at some of the more common types.

The plug-in converter is a small unit that allows you to convert any four-hole terminal to modular. They cost a dollar each and are worth every penny. Nothing to screw or rewire. Just plug it into any older four-hole terminal and you've joined the modular age. Of course, if you have a modular insert and you want to use your older phone with a four-prong plug, just buy a modular adapter, modular on one end, four-hole insert on the other. Plugs are also available that will allow you to add more than one modular piece of equipment to your line. These adapters, costing no more than three dollars each, are probably the most common type of plug. You can bet that as you add telephone equipment you're going to need several. Answering machines, for example, are usually placed near an existing phone. The phone is plugged in one socket and the answering machine in the other. As you'll discover in the section dealing with answering machines, however, some manufacturers have anticipated this minor inconvenience and have added a modular insert in the answering machine for your telephone. I would not recommend hooking your phone into the answering machine, however. Lose electricity and you lose both the service of your answering machine and your telephone. But how many times have you lost normal telephone service during a blackout?

TESTING YOUR NEW SYSTEM

Now that you have added new wires, replaced old equipment, and are ready to install your new phones and ancillary communications equipment, it is a good idea to run through a few of the basics to be sure you can both receive calls and place outgoing calls. Before checking actual wire connections, remember to disconnect your own premises' wire

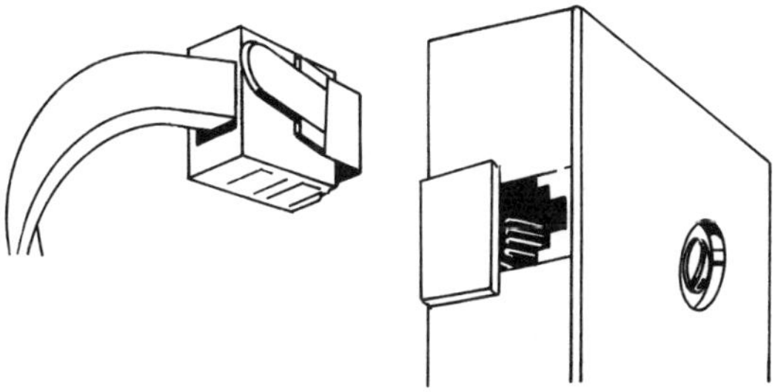

Modular Plug

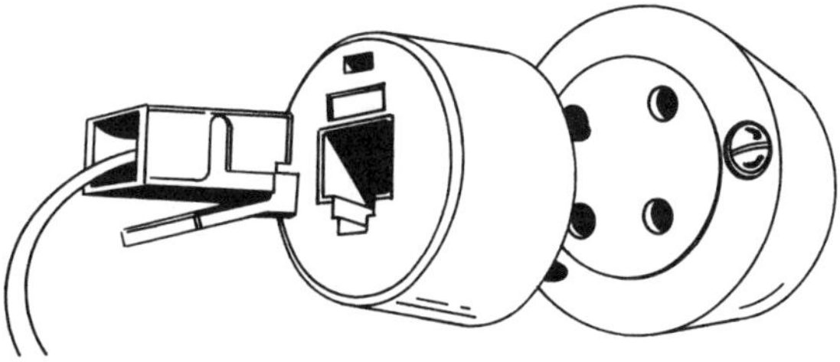

Four-prong to modular adapter

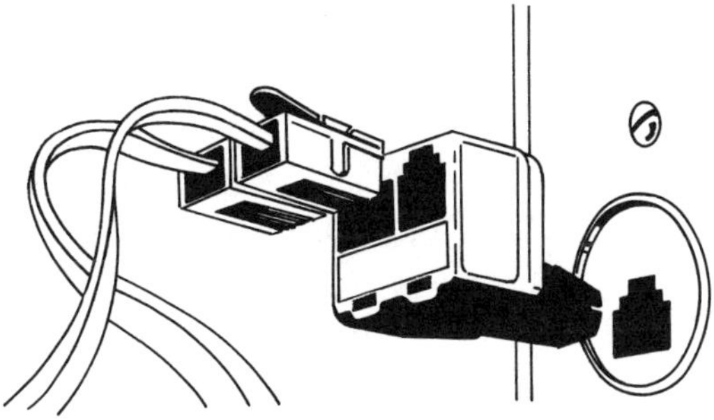

"T" adapter

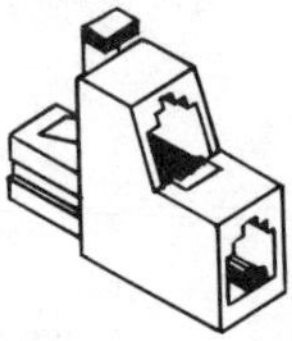

Dual Outlet Adapter

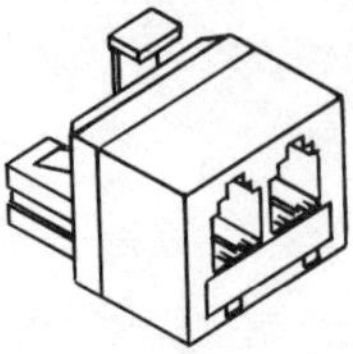

Dual Line Adapter

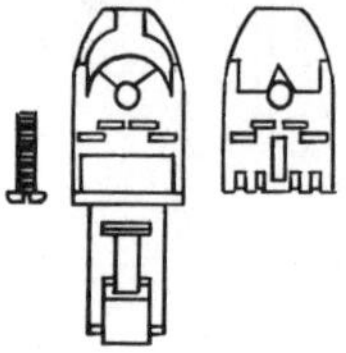

Telephone Wire Plug

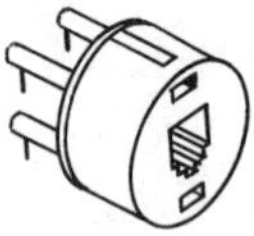

Permanent 4-Prong Adapter

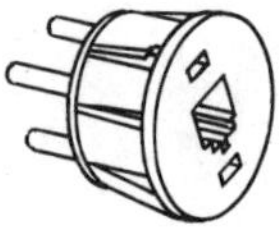

Portable 4-Prong Adapter

from the phone company's SNI. However, don't forget to reconnect the same wires to the SNI when you actually want to test your phones.

Placing Calls

1. Make sure you have a working telephone with a ringer.
2. Plug your working telephone into each newly installed jack.
3. Make sure you can receive a dial tone.
4. When you receive a dial tone dial a local number. You should be able to conduct a normal conversation without hearing difficulties by either party.

Receiving Calls

1. With the telephone still plugged in, have a friend give you a call.
2. Your new phone should ring. Again, you should be able to conduct a normal conversation without any hearing problems at either end.

TROUBLESHOOTING

If you run into a problem in completing the above tests, go through each link and locate the cause of the difficulty.

First, is it your problem or is the telephone company network at fault?

1. Start at the SNI. Unplug or unwire your connections.
2. Plug a working telephone into the telephone company's jack or attach a working adapter to the phone company's line.
3. Run through the testing procedures above for making and receiving telephone calls. If you don't have an understanding friend handy, call your telephone company's local office. They won't charge you for the call and, besides, think how pleased they will be helping you install your own equipment.
4. If you still have a problem, notify the telephone company's service department, listed in your local telephone directory. Describe to them the nature of the problem. Let them know that you have installed your own wires and equipment, and describe the testing procedures you have undertaken. The problem is now theirs, not yours.
5. If, however, the trouble disappears when you use a working phone from the SNI, I'm afraid it's your problem now.

If the problem is in the wiring, you'll have to locate the exact nature of the trouble before you can repair it.

1. Reattach you own connections to the SNI.
2. Trace the wire along the route from the SNI to your first junction box or terminal.
3. Test your phone as above.
4. If it works, trace your phone line, making a test at every connection point and jack. If it doesn't work, the problem is in your connection to the SNI or in the actual wiring of the nonfunctioning jack(s).
5. An easy way to determine if a jack is not working properly is to remove it and replace it with a jack that you know is working. If the problem persists, you'll have to further isolate the trouble by looking for possible problems with:
   —Broken wires.
   —Wires that are touching due to worn insulation.
   —Wire that has been damaged by its wire attachments.
   —Incorrectly wired jack or wire junction box connections.
   —Loose connections to jack or wire junction.
   —Incorrectly wired or loose interface plug connections.
   —Phone wires too close to other electrical wires or electrical apparatus (see page 52).

If you have Touch-Tone dialing equipment and can get a dial tone, but your telephone dial pad fails to work, try reversing the red and green conductors (wires).

If you still have a problem with the hookup of any of the modular parts, AT&T maintains a twenty-four-hour toll-free number to assist you. Call 800–222–3111. Their service people will even talk you through a problem if your phone lines are open. It is an extremely valuable and reliable aid.

When you have located the problem and have made the necessary repairs, you are all hooked up and ready to go. Now, what type of home communications equipment would you like to use?

Read on!

— 6 —

Buying Your New Telephones

Now that you've had a chance to add new wiring to your home, the time has come to talk about hardware—new telephones or extension units. The chances are that you have given a good deal of thought to the kind of phones or phone extensions that you want. You should know whether you are equipped to hook up phones with tone dialing or whether you need adapter plugs, whether you want cordless or corded phones, standard or designer phones, and the variety of options or, if you will, gadgets, that comes with so many newer models.

Also, give thought to price. Remember, it may be a truism that you get what you pay for; however, it's still a good rule to keep in mind. I reviewed and tested several dozen types and varieties of new phones and found, in general, that quality has a price tag. One inexpensive model of a GTE Flip-Phone dropped onto the floor while I was plugging the unit in. The Flip-Phone did just what its name implied—all over the floor. Another model, a twenty-dollar phone, lost the spring that held its cover in place. The manufacturer did stand behind the unit and sent me a new one. However, I was out postage and time and suffered a lot of inconvenience.

Give thought to the use you're planning to give the phone. If it is a bedroom unit, consider some of the designer models. Only a few companies, including AT&T and Krone, manufacture a wide range of colored phones. If you are going to use the phone in a kitchen or work area, beware of those small light models—particularly the phones with a light base. When Ma Bell first introduced the popular Princess phone she had first spent hundreds of thousands of dollars on design. The result was a beautiful instrument that eventually found its way into the Museum of Modern Art. But it had a serious drawback. The phone was so light that every time the user stretched the cord between the handset and base beyond a critical point, the base flew off the table, often whacking the user. The moral: Beware of phones with light bases and tight cords.

Let's review the types of telephones available:

✆ The *basic* telephone may be defined as a two-piece unit with a corded handset that usually offers only limited features. Phones in this category include the conventional electromechanical phones you have rented from the phone company.

✆ *One-piece* telephones generally combine all the electronic components into a single hand-held unit. Most one-piece phones are typically compact and lightweight and allow you to move around easily while talking.

✆ *Multi-function* telephones are fancy electronic devices that offer you a wide range of communications features such as memory, redial, mute, and the like. (See page 43 for a description of features available.)

✆ *Decorator* telephones typically emphasize a distinctive decorative appearance. They include nostalgic types such as European cradle phones, Early American candlestick phones, character phones such as Mickey Mouse or an Olympic hero, chest phones, gum ball machines, etc. These phones can certainly complement a room as well as add novelty. However, keep in mind the fact that styles change. If you don't mind discarding your phone next time you redecorate a room, there are certainly some interesting phones from which to select.

✆ *Cordless* telephones can operate without a cord because they use radio frequencies for communication between the remote handset and the base unit. The base unit is still, however, manually connected to the telephone network. Cordless phones may include an intercom feature that allows conversations between the base and remote handset. Some cordless telephones also provide a security feature that prevents other cordless users from using your phone lines even though they have the same frequency. With no telephone cord to restrict your use, cordless phones provide tremendous mobility. While talking on the phone you can roam from room to room, relax by the pool or in the garden, or visit a neighbor—as long as you stay within a three-hundred- to one-thousand-foot radius of the base unit. (Short-range cordless phones can usually extend no more than one hundred feet from the base.) The use of radio frequencies leaves cordless phones susceptible to interference from electrical noise, building structures, and often nearby cordless-phone users sharing the same frequency. The FCC recently assigned a total of twenty frequencies, or ten channels (two frequencies are required for each channel pair), in the 46/49-MHz range to help remedy interference problems.

A word about cost. If you want a cheapie, there are more than enough throwaway telephones on the market that can be bought for as little as five dollars. And that's about what they are worth. The phones described in this section all meet a basic quality standard and are warranted by the manufacturer. The prices mentioned below are the list price or the manufacturer's suggested retail price. In many cases you can buy your new phone for less than the mentioned price. Major national chains, such as Sears, Caldor, Radio Shack, and so on, have sales with list prices cut as much as a third. Check them out. Some of the more specialized phones described may be available only at phone stores or from several of the major reliable catalog shops such as The Sharper Image in California. If you can't locate a listed telephone model in your area, the directory of phone manufacturers and suppliers at the back of this book will give you addresses and, where available, a toll-free number.

Now let's take a look at the various models currently available. (Note: I have eliminated many individual models and many phone manufacturers that were either below standard or were not easily available.)

BASIC TELEPHONES

Radio Shack

Perhaps the least expensive basic phone is manufactured by the Tandy Corporation and is sold through Radio Shack. It is the Model 43–302. The phone is white; however, it also comes in beige (Model 43–301). The price is $39.95. The unit is a standard rotary-dial phone that will plug into any modular outlet. It comes with a seven-foot base cord and a bell volume control. This can be considered a stripped-down model of a solid, reliable telephone. For another twenty dollars, however, Radio Shack has a standard desk Tone-Fone unit that also comes in beige (Model 43–341) or white (Model 43–339). Unlike the rotary-dialed unit, this unit can be used with any of the long-distance services. It also comes with a detachable four-foot handset cord. Want a longer cord? Unplug the old, plug in the new.

Webcor

A slightly sleeker model of the familiar desk phone is Webcor's Model 1000, priced at $29.95. This model features a push-button keypad that produces pulses and operates on rotary- or tone- (but not true tone) dialing systems. It also features a last-number redial button that stores the last number you dialed. If you are planning to use a desk-style phone with any of the long-lines services, the above model won't do. Instead, look at Webcor's Model 1000ZT, which has a true tone keypad. Both models come in beige, brown, or white and have bell volume control buttons.

ITT

ITT produces a desk model unit (Model PC2400) that has true tone dialing plus last-number redial and a ten-number memory capacity. Although it doesn't have a mute switch or a switch that can turn off the ringer, it does contain an adjustable volume control. The unit comes in white or brown and retails for $110.00—one of the highest prices of the group.

GTE

GTE also makes a basic desk phone called the Fashion Plate, which comes with thirteen different colored faceplates so you can mix and match the colors to harmonize with any decor. The phone is available with a rotary and a tone dial, or what GTE calls its Universal Dialing Keyset. This means push button on the outside, rotary on the inside, but no true touch-tone—in other words, it looks push button but won't work with the long-lines services. No fancy features. The price is $54.95. GTE also manufactures a more contemporary two-piece desk phone called the Ultraline II. Fully push button, the unit is equipped with an automatic redial, a nine-number memory, and a three-position electronic tone ringer. The retail price is $69.95.

Mura

Mura also produces a line of standard desk phones. Its models vary in price depending on the features you select. You can have a regular rotary-compatible model that lists for about $29.95. Mura's standard desk phones differ from most of the basic models in that they begin to add some interesting features as part of the package. Its Masterhold addition (on Model DP 463), for example, gives every phone on the same single line system independent hold regardless of brand or model. Answer a call in the kitchen or den, put the caller on hold with one touch of the hook switch, and replace the handset. You're now free to pick up any extension and resume your call without stray noise or less volume. This particular phone stores up to ten numbers in its memory (up to sixteen digits) and dials automatically with the push of two buttons. The unit also has a mute button. The Masterhold phone sells at The Sharper Image for $79.00. All Mura's desk phones have a ten-number memory plus a last-number redial feature. Available colors are white, almond, and brown.

AT&T

Ma Bell, of course, pioneered the basic desk-model phone. To eliminate this model from the line would be like McDonald's giving up its Big Mac. This phone is the heaviest of the lot, weighing a good 25 percent more than its competitors. The price varies depending on where you buy it. Several discount stores now carry the AT&T line. Sears has sold the basic rotary model for $32.99; it lists for $39.99. The Touch-Tone model lists for $61.99 but sells at Sears for $49.99. These basic phones are also available in wall models, with rotary or Touch-Tone dialing. AT&T provides the widest assortment of decorator colors: white, ivory, black, yellow, and light blue. The desk model has a six-foot coiled cord from the base to the handset; the wall model's cord is twice as long.

PRINCESS-TYPE PHONES

One of Ma Bell's major design innovations during the late 1950s was the introduction of the small bedside Princess phone. Following on the heels of the Princess, AT&T came out with its Trimline phone, which kept the same basic design as the Princess but incorporated the dial pad in the handset rather than the base. Having the dial and handset together was not an AT&T invention, however; the Swedes started it several years earlier with their Ericaphone.

AT&T

Ma Bell still sells her Princess and Trimline phones by the multithousands. For many it is the design as well as the tone-quality standard. The basic Princess lists for $64.95 and is available in the basic Bell colors: white, ivory, yellow, and light blue. The unit is still available with either rotary or Touch-Tone dialing and is made only as a desk unit. The Trimline, which has the dial pad in the handset, is the unit that Bell hopes will eventually replace the Princess. It is smaller and lighter. (The Trimline weighs only 2½ pounds; the Princess is 5 ounces heavier.) The Trimline is also available in additional colors—brown and rust. Choose between a desk or a wall unit. The Trimline lists for $69.95. Since neither the Princess nor the Trimline comes with the host of added gadgets Bell's competitors have added, Ma recently introduced a new line called the Touch-a-Matic 300 series. The basic phone in the line is the Model 300/305, which lists for $79.95. It is an attractive, boxy-looking modern phone that comes either as a desk or a wall model. As a desk unit, however, its light weight, at 1½ pounds (the wall unit is 2 ounces less), is a disadvantage. It is quite easy to pull it to the floor despite its seven-foot coiled cord. Both units are dialed from the handset and are push button only; either has pulse/tone switchable dialing (if your local phone company isn't equipped to handle the tones). You can program two numbers of eleven digits or less, and can recall the last number dialed. It comes with a mute button and a three-level ringer switch. If you want a handsome desk phone that is fully modular and are not concerned with price, look at Ma Bell's Director telephone, with its solid rectangular block of walnut-stained cherry wood as the base unit and a Trimline handset. The unit comes with only a single colored handset, black. It has the same features as the standard Trimline phone. The price is $99.95. AT&T gives you a one-year warranty on all its phones.

Radio Shack

Tandy produces several inexpensive phones that are Trimline look-alikes. Its Model ET–270 can be used as a desk model or it can be wall mounted. It is available either with pulse dialing or with true tone dialing. However, only the pulse model has a last-number redial and a mute switch feature. The cost is $39.95. For another twenty dollars you can buy Radio Shack's Model 43–355 or 43–351 (desk model only), which do essentially what the lower-priced unit does but look more like AT&T's Trimline phones. Tandy calls its line Trim-Fone. You have a choice of two colors: white or brown. Radio Shack's warranty is only three months.

GTE

GTE makes a series of slimline phones it calls its Solitare series. The basic Solitare is an inexpensive device, at $14.95, that is only an answering unit—it has no dial, no outgoing calls. GTE's Solitare IV also has a low list price, $29.95, and can be used as a desk phone or be wall mounted. Its push buttons can be used with rotary service; however, it doesn't transmit the tone signals needed to operate the long-lines services. Other features include a mute switch and both a ringer and a phone on-off switch. In January 1984, GTE introduced its Solitare VI, which can be set to dial calls by either the true tone dialing or the pulse dialing method. Just flick a switch. This phone stores nine numbers in its memory and automatically remembers the last number dialed. It comes in one color, almond brown. The list price is $54.95. The warranty is for three months.

GTE also manufactures one of the most-imitated telephone instruments, the Flip-Phone. This instrument has a plastic shield that covers the dial pad on the handset. There is no cradle. Just lift up the handset and the cover "flips" open, revealing the dial pad. Curiously, there is no visible mouthpiece; it is hidden along the bottom edge of the phone under the plastic shield. GTE's newest, the Flip-Phone II, was introduced in 1983 and is basically the same as its previous model except it has a last-number redial feature. Its push-button dial pad works only with pulse-dialing systems, which is a disadvantage for those using the long-lines services. This lightweight phone also has a mute switch and a ringer volume control switch. It comes in white, almond, brown, and rust. The price is $54.95.

Teleconcepts

An interesting new twist in home phones is the Thinfone, introduced by Teleconcepts. It's the first Trimline phone with the ability to handle conference calls. You can put one call on hold, dial a different number, and, when connected, merge the lines for a three-way conversation. Far less expensive than business phones, this unit gives you lighted indicators for "line ringing," "line in use," and "hold." It also has an illuminated keypad for night dialing, a mute button, a ringer volume control switch, and, in the pulse mode, an automatic redial button. The colors are brown or ivory. The price of the wall-mounted phone is $89.95; the desk model lists for $79.95.

Teleconcepts also distributes a smaller, less-expensive phone called the Cadet. It can be used as either a desk or wall phone, comes in brown or ivory, and provides either tone or pulse dialing. The standard pulse- or tone-dialed phone is $29.95. Add a pulse/tone switch, and you add another ten dollars to the price. Throw in a ten-digit memory feature, and the price is upped another ten dollars.

Webcor

Webcor also makes a reliable, inexpensive one-piece phone (Model 720). Its basic unit, which operates with either rotary or tone (but not true touch-tone) dialing, lists for only $14.95—really a disposable phone. And you get surprisingly many options for your money: a last-number memory redial, a mute button, and a ringer volume control switch. If you want to use tone dialing, Model 720ZT will give it to you for another ten dollars. Webcor also makes more sophisticated units, for more money, that give you more features. Their Model 708, for example, is a more streamlined unit than the less-expensive model. It gives you true tone dialing, for MCI, SBS Skyline, and the other long-lines services. Its colors are beige or brown. The warranty is for three months.

Mura

Mura has a complete line of Trimline-style phones in a wider range of colors than most suppliers. Its designer phone, Model CV 444, is available in white, almond, brown, red, gray, or blue. It has switchable pulse/tone dialing, a ten-number memory with last-number redial, and an adjustable ringer. The list price is $49.95. Mura's newest in the line was introduced in the summer of 1984 and incorporated its

Masterhold concept. Install just one of these new phones, and with a brief tap on the unit's hang-up hook, every phone in the house goes to hold until the call is picked up again at any extension. Since each phone shares the Masterhold feature, the incoming call does not have to be received on the specific Mura phone. After two minutes on hold, if not picked up elsewhere in the home, the phone is preprogrammed to release the call. This phone, Model CV 464, retails for $69.95. It also comes with pulse/tone switchable dialing, a ten-number memory, last-number redial, an adjustable ringer, and is available for either desk or wall use. Its colors are brown, almond, or white.

One problem with these phones, I have discovered, is the design of the base. It is very easy to replace the handset on the cradle backward. The phone seems to fit nicely and you may believe your connection is terminated. Wrong! In a few minutes you'll hear that awful buzz that tells you a phone is off the hook. Miss this sound, and all the phones in the house are out until you discover the problem.

Mura also makes several series of one-piece phones that sell from $14.95 up. Its Model GK 200, which lists for $14.95, is a basic wall or desk phone, whereas the Model GK 350, with a rather high REN of 1.9, has a ten-number memory, last-number redial, a mute switch, a ringer volume control switch, and is priced at $39.95. All Mura's phones come with a one-year warranty.

General Electric

GE has an interesting phone it calls the Hotline. It's a new twist on the memory concept. Their phone gives you instant access to emergency services with the three specially marked one-touch buttons. The recessed buttons are also illuminated to make help at night one push away. In addition, the unit has nine more standard memory buttons (up to sixteen digits each), auto redial, a mute button, a ringer volume control button, and an extra-long, fifteen-foot coiled cord. The phone works with a rotary or tone line, but it doesn't give you true tone for the long-distance services. The price is $59.00, and all GE's telephones come with one of the longest warranties in the field, two years.

Cobra

Cobra introduced its newest series of slimline phones in mid-1984. Its Series 400 line includes lighted touch buttons and on-off ringer buttons. The least-expensive unit, Model ST–401, at $24.95, gives you a basic pulse-dial pad with a last-number redial button. Cobra's premium model, ST–410, includes a twelve-number memory, pulse/tone switchable dialing, a mute button, and last-number redial. It comes in white, beige, or brown.

Krone

Krone manufactures one of the most attractive telephones on the market. Its Compact 1000 series was designed by Mario Buatta. This is a two-toned phone with a charcoal base unit and a handset available in white, sand, gray, red, blue, or black. The phone lists for $130.00. It is designed to be used with any Touch-Tone line and comes with a volume control and mute switch. An extra-long sixteen-foot coiled cord is standard; a twenty-two-foot cord is an available option. Krone, a West German company, has been in the phone business for more than half a century and gives the longest warranty around, five years.

Panasonic

Panasonic also manufactures a slim-type phone that works with any pulse dialing system. Its newest, Model KX–T2200, lists for $39.95 and is available only in white. It gives you a last-number redial, a mute button, a three-level ringer volume control, and can be used as a desk or wall model. The warranty is for a year.

CRADLE PHONES

The cradle phone has come a long way since you had an exposed hook on which to hang the earpiece after completing your call. Ironically, these early instruments are now collector's items and are being reproduced for those who want a modern modular phone with an old-fashioned look. Ma Bell has combined modern technology with a flair to capture the antique appearance of the old phones with two models. The Country Junction is a look-alike of the 1913 original. It has a separate wooden handset that is connected to the wall unit by a straight thirty-two-inch fabric-wrapped cord. The wall unit has the old nickel-finish rotary dial and a pair of bells. When was the last time you heard a real phone bell? If you would rather have a desk model, take a look at the Country Squire. Again, AT&T has copied the old handset and added a braided fabric cord; instead of a rotary dial, a circular Touch-Tone dial gives you the look of old with the convenience of push-button true tone dialing. AT&T has even copied the antique keyhole in both models to maintain the early look.

If you like the old-fashioned look but prefer the European style, GTE makes three different European look-alikes. Its Sweet Talk is a French-style cradle phone with a rotary dial; its Cutie phone is smaller than the French-style with rich gold trim. Both sell for $59.95. GTE also produces an English-style rotary phone that lists for $69.95. All the GTE models are available only in ivory; all are completely modular.

Radio Shack has a line of European-style phones ranging from the French-style at $59.95 to the smaller Cutie at $49.95. It also sells its Imperial Classique, which looks like it was just pulled from the wall of a Viennese boudoir; the cost is $79.95. Its newest phone is a replica of the old

American candlestick design called the Bonnie 'n Clyde, which is black with gold trim. It sells for $59.95.

Teleconcepts also distributes the old-fashioned look. Its French phone has a rotary dial and comes in white, ivory, or gold and lists for $79.95 (the gold model is ten dollars more). It has one of the most ornate instruments on the market, which it calls its Persian phone. The gold base has an etched appearance with the thin handset right out of an old Peter Lorre spy movie. It also retails for $79.95. Teleconcept's candlestick phone is called the Speakeasy and comes with a plastic base (in black, white, red, ivory, or black with gold trim) for $59.95, or a gold metal-based phone for $79.95.

An extremely impressive modern European telephone is the Danish phone that won the coveted Danish Industrial Design Award and is on permanent exhibit at New York's Museum of Moden Art. It works only on tone-dialing systems and comes with a solid black handset and a choice of four colors for the base: white, red, gold, or brown. It is available from The Sharper Image with a one-year total replacement warranty for $129.00.

The more popular cradle phones give you more than just a hook and a receiver. The geegaws such as memory, last-number redial, and mute and ringer switches, are most common. All the phones vary in appearance and try to capture what the manufacturer believes is the going trend in consumer appeal. They range from the relatively simple units to complete systems.

Another feature that appears to be popular is the big-button phone. AT&T manufactures a unit that looks like a box filled with children's playing blocks. Each large block has a number or symbol. A modern handset and its cradle are on the side of the box. Totally push button, pulse dialing and a ringer on-off switch make this phone adaptable for a young child's room—if you have children who know how to use a phone. The phone sells for more than $150.00. A less-expensive look-alike is made by Webcor, called the

Webcor Electronics, Inc.

Jumbo Button phone. It looks more like a modern sculpture than a box of blocks. It also has features Ma Bell excluded: pulse/tone switchable dialing, last-number memory redial, and a mute button. It is available from Webcor or The Sharper Image for $89.00.

Let's take a look at the range of other cradle-type phones that are available:

GTE makes two lines of two-piece phones that can be wall mounted or used on a desk: the UltraLine and the Linear. The basic UltraLine is a sleek, modern phone that is available with either pulse or tone dialing. Only the pulse unit, however, gives you last-number redial. The price is $59.95. For another ten dollars you can have the next model, the UltraLine II, which gives you the last-number redial feature as well as a ten-number storage capacity. If you want the option to switch between pulse and tone, look at the Ultrastar, which sells for $74.95. All these phones are available only in white except the Ultrastar, which is also available in almond.

GTE's Linear phones are also switchable between pulse and dial and have a ten-number memory. However, they have added a new feature—an electronic call hold that allows the user to place a call on hold and pick it up on another phone. An indicator light near the hold button lights up when a call is being held. The price is $109.95. Want to go whole hog? GTE has a phone it calls the Gamma, which has all of the above features plus a few more: a memory that stores not ten but sixty-two numbers of fifteen digits each, a clock, an alarm, a timer, and a send switch—to dial automatically a number being displayed on a liquid crystal display. It comes in a wall or desk model, and costs $179.95.

Teleconcepts distributes several interesting two-piece telephones. Their Perkie desk or wall phones are very modern in appearance with an emphasis on curves. Both have a pulse/tone switchable push-button illuminated keyboard, a redial button, and a ringer volume control switch. Colors are taupe and ivory; the price, $59.95. If you are interested in the square look, the company also has its Lotus phone. Here is where advance design using microchip technology really becomes apparent. The Lotus is less than an inch thick, with rounded, soft touch buttons arrayed in two linear columns. The electronics, colors, and price are identical to the Perkie. It's also available from The Sharper Image.

Cobra is best known for its cordless phones, but it also manufactures an interesting corded phone that does more tricks than most of the phones on the market. Its Model ST–660, which lists for $99.95, features a multi-functional digital LED display for time/day of week, call timer, alarm system, and dialed number confirmation. It also has an electronic phone book, or visually displayed memory for storing fifty-eight numbers of sixteen digits each. It has an adjustable ringer as well as a switchable pulse/tone dial button. Cobra also makes a two-line version of this phone for $129.95. Not only does that phone, Model ST–662, have all the features of its sister unit, but it also has separate ringers for each line plus an electronic hold and three-way conference switch.

Webcor also makes a series that is available in either the single- or dual-line mode. Model 818SW is its single-line pulse/tone switchable phone that allows you to store thirteen numbers. It has the same features as the Cobra phone, but lists for less, $83.95. Webcor's two-line phone (requiring two separate incoming phone lines with two telephone numbers), Model 828SW, is $109.95. Both phones can be used as either a wall-mounted or desk phone. Colors are beige, brown, or white.

Uniden is best known as a manufacturer of high-quality cordless phones. However, in early 1985 it introduced a corded phone that appears to do about everything. Model EX 1070, listing for $269.95, is an ultra-thin handset containing a large memory auto-dialer, a synthesized voice recorder to act as an answering machine when you're out, and all the other features that set this phone apart.

ADD A SPEAKER TO YOUR TELEPHONE

If you want to go one step above a regular gadget-laden telephone, think about a phone that allows you to carry on a two-way conversation without holding a receiver or speaking into a hand-held device. Although many consider the speaker phone just another gimmick that manufacturers have added to the grab bag of telephone options to entice you, in a few instances it serves an important need—for business conferences, for the individual with too many things in his or her hands and no room for a phone, or for the handicapped. However, in most cases the disadvantages of speaker phones do not compensate for their value. The primary problem with speaker phones, at least from my point of view, is acoustical. If you have ever listened to someone talking over a speaker phone, you know what I mean. Rather than coming across in a clear, sharp voice, your phone companion sounds hollow. It's like listening through a pipe. Also, the speaker phone picks up all background sounds, such as paper being shuffled, a door being opened, or the stirring of a pot. If you're convinced that you want a speaker phone, you have quite a few choices. And you don't have to run out and buy a new unit; your present phone can easily be altered by the addition of an amplification unit. Before looking at the available speaker phones on the market, here are a few adapters that are extremely easy to install and relatively inexpensive:

Radio Shack provides the least-expensive speaker. It looks like a small white box (6 inches long by 4 inches high) with a few buttons on the top. To operate it, just plug its five-foot modular cord into any modular telephone jack. To speak, hold down a button; to listen, release the button. The cost is $29.95. For a better unit both in terms of features and tonal quality, spend another twenty dollars for Radio Shack's DuoFone–101. It has an automatic talk/listen switch, a mute button, and a slide-action volume control. Rather than boxlike, this is a flat, two-inch-high desk unit measuring approximately 7 by 7 inches. Both units are available only in white.

Webcor also manufactures a reliable, inexpensive speaker phone attachment (Model A–60) that measures 7½ by 5½ inches. It is designed to be either a desk or wall-mounted unit. After a number is dialed, simply press the "talk" button, return the handset to the cradle, and carry on a two-way conversation. An LED light is activated when the talk button is in use. To disconnect, press the "talk" button twice. The colors are beige or brown. The cost is $38.95.

AT&T also manufactures a basic speaker attachment for $64.95 (Model 600). This unit, however, requires you to plug the modular speaker cord into a phone jack and plug your telephone's modular plug into the back of the speaker phone. It also has an adjustable volume control and a mute button. A small attractive unit measuring roughly 6 by 6 inches, Ma Bell's high-quality speaker phone adapter is designed to be either desk or wall mounted.

PHONES *WITH* SPEAKERS

If you are about to buy a new telephone and want the speaker feature, don't consider the separate speaker unit. There are enough quality phones on the market that give you a speaker option that it simply doesn't pay to mix and match. Telephones with speakers built into the instrument are readily available on many corded phones and most cordless phones.

Webcor, for example, produces a reasonably priced line of multi-functional phones that combine the speaker feature with sixteen-digit auto-dialing. It also provides an interesting feature: an automatic redial of the last number called, every minute for up to fifteen minutes—a good way to solve the problem of a constant busy signal if you don't want to keep redialing and waiting. These Webcor phones also have the usual geegaws—pulse/tone dialing, hold, an on-off ringer switch, plus a blinking LED. Prices range from $79.95 for Model 757 to $119.95 for Model 759. They all come in ivory; however, if you would like a wood-grain finish, add about seven dollars. These multi-function phones also have a two-line option for an additional seventy dollars. They are among the least expensive multi-functional phones that provide quality. Again, you can buy cheaper speaker phones, but listen to one before writing your check; you may change your mind.

Mura also makes a very fine quality speaker phone series that is comparable in price to Webcor's. Mura's FP 200, for example, is quite a bit sleeker than Webcor's phones and comes with a white handset that rests on a black base unit. It has a lighted keypad, a ten-number memory, and up to nineteen digits per number. The cost is $79.95. For ten dollars less, however, Mura produces the same basic features as the FP 200 without the contemporary appearance. The FP 50 is a standard white push-button unit that can be either placed on a desk or wall mounted. If a ten-number memory isn't enough for you, look at the FP 100. It is similar in appearance to the FP 200 but costs $89.95. For the extra expenditure you can have a twenty-five-number memory, last-number redial, a mute button, and a battery back-up to protect the numbers in your memory chamber in case of an electrical failure.

Panasonic is not to be left at the gate as the various phone manufacturers vie with each other to provide you with more gadgets and fancier features. In late 1984 it introduced its KX–T2340 speaker phone at a list price of $89.95. This phone gives you a thirty-number memory with up to thirty digits per number. The difference in the numbers of digits that can be held in the memory bank can mean pushing one button or two to use the long-lines services, certainly not a major problem. Panasonic, however, has added three additional buttons, which it refers to as "direct-dial, on/off hook dialing." The idea is to store three emergency numbers of your choosing—police, fire, ambulance, for example. To reach any one of these emergency numbers, you have only to press one button, wait for the call to be completed, then talk directly into the speaker phone; there's no need to remove the receiver from its cradle.

Radio Shack has a speaker phone, DuoFone–160, that retails for $139.95 with a switchable pulse/tone feature plus a thirty-two number, twenty-eight-digits-per-number memory auto-dialer. The phone is available only in white and, unless you can buy it on sale, is overpriced.

TELEPHONE HEADSETS

Want to play operator? No need to hold on to a handset or talk into a speaker. Telephone operators, and NASA astronauts for that matter, have been using earphones and speaking into a specially attached microphone for years. The

Elite, sold by The Sharper Image, gives you the same hands-free convenience as you talk, with a ten-foot cord for freedom of movement. The headset plugs into the base of your phone in place of the handset. To make or answer a call, just take the handset off the cradle, as you normally would. The 3½-ounce headset rests on the outside of your ear with a replaceable soft foam ear cushion. The amplified receiver has an adjustable volume control. The cost is $59.00. The deluxe version, with coiled cord and quick disconnect feature, is $99.00. Radio Shack also sells a telephone headset (Model 43–169) with regular and small foam earpieces, an adjustable mike boom, and a six-foot cord with "clothes clip," for $69.95.

CORDLESS PHONES

One of the latest developments in telephone technology has been the introduction and wide acceptance of cordless phones. Actually, the cordless phone is a small radio receiver and transmitter. The receiver is contained in the phone's base unit, which is plugged into a conventional telephone line by means of a standard modular plug. The telephone/radio's transmitter is powered by plugging another cord into a regular electric socket. The radio transmitter/receiver in the portable handset is powered by a sealed battery within the handset. The battery will hold enough power to allow you to separate the two units for about twenty-four hours. Then you must replace the handset in the base unit to recharge the battery.

One of the problems that can occur with cordless phones is a power failure. Lose your house current, and your cordless phone won't work. Another problem is one of interference with other cordless-phone users in your area. The FCC has allocated only a limited number of frequencies over which cordless phones can transmit radio signals. (In March 1984 it increased the number of channels available to telephone communications to ten, with plans to double that number in the near future.) It's only logical that in any congested area with many people speaking over their cordless phones circuits will overlap. Also, since cordless phones transmit over a specific set of radio frequencies, 49 MHz, anyone with a shortwave radio can tune in on your calls. The older cordlesses transmit on 1.7 MHz or 1700 kHz, which is just above the 1600 mark on your AM radio. People with sensitive radios have complained that they can pick up telephone signals at the top of their dial.

A more serious problem is the unauthorized use of your phone lines. With most cordless phones, nobody can transmit—use the phone—when the handset is being recharged on the base unit. Remove the handset, however, and your line may be fair game. Anyone with a cordless phone having the identical radio frequency as your phone may be able to make calls over your phone line. Some of the newer cordless phones have added owner-selectable frequencies and digital security coding to try and frustrate unauthorized use. However, since all cordless phones are broadcasting over a specific radio frequency, it is almost impossible to be assured that your phone line won't be used when your handset is removed from the base unit.

The distance your handset can operate from the base unit is based on two factors: the power of the system and obstacles between the handset and the base. Within fifty feet of the base, most cordless phones operate without the need of a handset antenna; the maximum distance a cordless phone will operate effectively, with its antenna, is about one thousand feet. Obviously, some of the lower-powered, less-expensive units will only work effectively within five hundred feet or less. The quality of voice reception and transmission over a cordless phone is rarely as good as a corded phone. However, with most of the better-quality cordless phones, the reception is certainly better than adequate.

In selecting a cordless, keep several factors in mind: for one, your base unit—the boxlike device upon which the cordless phone *must* rest when not in use for any long period of time—is not the most attractive accessory. It has a twenty-six-inch aerial that must be fully extended if you want to transmit phone signals the maximum distance. Keep in mind that you need to place your phone's base near an electric outlet. Also, when shopping for a cordless phone read the warranty that accompanies your phone. Be sure there is a service center near you in case of problems. The handset must use a battery. And although nickel-cadmium batteries are quite efficient, eventually they will require replacement.

Perhaps the most serious drawback with some of the lower-priced cordless phones is their inability to be converted to true tone dialing. Thus, you can't make outgoing calls using any of the long-distance services like Sprint, MCI, and SBS Skyline. One advantage with most cordless phones, on the other hand, is their ability to serve as an intercom between the portable handset and the base unit. By flicking a switch on the base, your system automatically converts to a two-way walkie-talkie.

Most cordless bases have a modular socket as a standard feature. This allows you to plug in a regular corded phone if you choose—an important asset if the handset is not near the base unit when a call comes in. Rather than run amok chasing the sound of the bell, just answer your corded phone in the traditional fashion.

There is a wide variety of reliable cordless phones on the market. AT&T sells a line of cordless phones it calls its Nomad series. All AT&T cordless phones have dial pads that work with both pulse and Touch-Tone service, a redial button, and a mute button. Although these phones operate with Touch-Tone service, they don't all provide the true tones necessary for Sprint, MCI, and other long-distance services. The lowest-priced phone in the line, Model 2000 ($99.95), is designed for use within one room—it has a range of but fifty feet; however, with this short range no antennae are required on either the base or the handset.

This is an attractive phone in white with gray accent. It cannot, however, be used as an intercom. For $189.00 you can buy AT&T's Nomad 600, which is a pocket-sized handset and a neatly designed base unit. This phone has a seven-hundred-foot range, a two-way paging/intercom function, and a security feature. Its only color is white. The warranty is for one year.

Uniden is perhaps the best-known maker of cordless phones, although it also makes both corded phones and answering machines. The heavy promotion of its cordless models with Jack Nicklaus as its TV spokesman may be familiar to many. In mid-1984, Uniden introduced a new line of cordless phones, its 49/49 series, designed to improve and upgrade its older models and, more important, to take advantage of the newer frequencies—46 to 49 MHz—that the FCC has allocated to cordless-phone manufacturers to try to prevent overcrowding the airwaves with telephone signals. Most Uniden cordless phones have a range of one thousand feet and some type of security device: either an option to select separate wavelengths or a simple cutoff device in the base unit so that no one can make calls on your line when the handset is in its recharge position. The Uniden EX 2100 is a basic unit without an intercom or a pulse/tone switch. It has a redial and a mute switch. The list price is $119.95; the color, white. To compete more directly with AT&T's smaller pocket handset, Uniden announced the EX 2700, which has the same basic features as the 2100 except that it can be folded much like the GTE Flip-Phone and it has a one-way page button. It can send a tone from the base unit to the wandering handset but you can't "talk" back. Its price is $159.95; its color, white. If you want something that gives you more—for more money, of course—the EX 4100, at $199.95, has both a safety feature when the handset is in its base plus a selection of sixteen selectable frequency codes. It also has redial, a two-way intercom, and a speaker phone. The most-expensive Uniden, which gives you a switch for pulse or tone dialing, is Model EX 4800, at $279.95. In addition, you get a few more features: an intercom, a speaker phone, 256 digital security codes, a memory of nine numbers, and a replaceable nickel-cadmium battery pack. All Uniden cordless telephones have a one-year warranty.

Cobra is another manufacturer well known for its cordless phones. In the summer of 1984 Cobra also introduced a new line featuring the 46/49 MHz frequencies. Like Uniden, Cobra phones contain consumer-selectable frequencies and digital security coding. Its premium model, CP 460S, retails for $199.95 and comes with two consumer-switchable channels and 512 digital security coding combinations plus a speaker phone, a nine-number memory, a redial, a hold, an intercom, and switchable pulse/tone dialing. If you don't need the speaker phone, Model CP 450S gives you everything the 460S does, except the speaker, for twenty dollars less. If you can do without tone dialing, the CP 440S has pulse dialing only; it also has thirty-two selectable digital security codes. The price is $139.00. Cobra also has mini-cordless phones that have the same basic features as its larger look-alikes. Feature for feature, the minis cost about twenty dollars more. The colors are platinum-gray handset and base contrasted with charcoal-gray control panels. The warranty is for one year.

Mura, the company that initiated the petition to the FCC in 1982 to increase the number of cordless-telephone frequencies, has a series of five cordless phones all designed to use the new 46/49 MHz frequencies. Its midpriced phone, Model MP–680, at $169.95, has a range of seven hundred feet, with two-direction security coding, enabling the user to program two separate codes—one from handset to base, another from base to handset—to virtually eliminate "dial tone snatching" by outside handsets or false ringing from the base. The top of the line, MP–900, at $249.95, has a thirty-three-number memory and the ability to handle five separate remote units. It also uses voice-synthesized speech to aid the user when storing numbers in the memory by

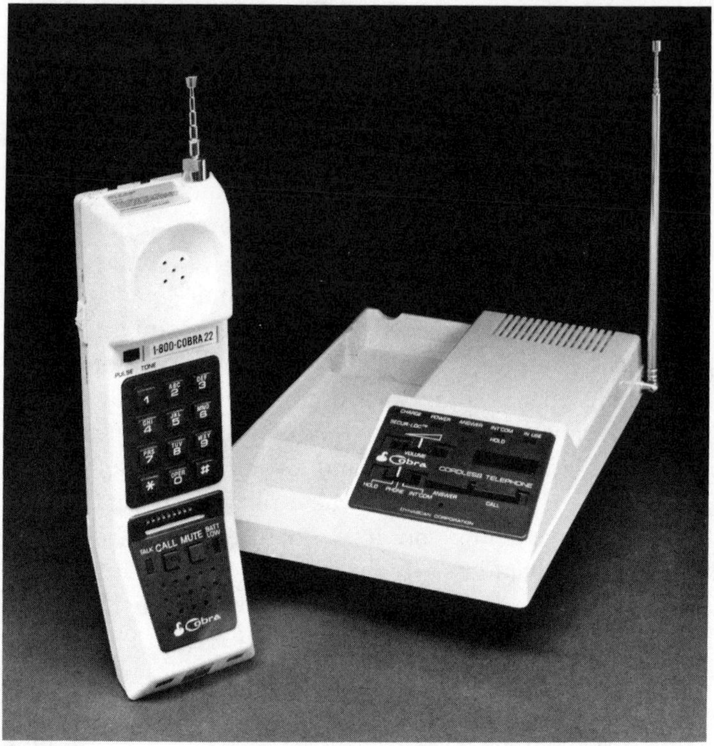

Cobra Communications

clearly calling out the numbers as they are being stored and indicating when programming is complete. The voice synthesizer is engineered to alert the user when a mistake is being made. Mura also sells a basic cordless, Model MP–450N, for $99.99, with many of the features of other companies' more expensive models.

Radio Shack has several cordless phones that have been in its line for a few years. If you have a 1983 catalog you will find them listed, but ignore the prices. The competition among phone makers has seen price reductions across the board. Radio Shack has cut its prices on cordless models by up to 20 percent. However, even with the price rollback it still lacks all the extras you will find on the Phone-Mates,

Unidens, and Cobras. Radio Shack's least-expensive two-way model, ET–320, lists for $119.95. It has a range of but one hundred feet and lacks true tone dialing for use with Sprint, MCI, and the other long-lines services. It does have the auto-redial and mute features. For the same price Radio Shack offers its DuoFone ET–340, which has a range of seven hundred feet but lacks the ability to make calls; it is simply a receiver. Both units are small, compact, and pocket-size. If you don't mind the lack of a dial pad on the handset, Radio Shack has an inexpensive mini, Model ET–350, designed for use within a small area. It has a range of one hundred feet with the dial pad on the wall or as a desk mountable base unit. The price is $79.95. The top of the line, Model ET–330, at $179.95 has a five-hundred-foot range, auto-redial, a security system, and a mute button. A major disadvantage with all of Radio Shack's cordlesses is their lack of true tone dialing for use with the long-lines services.

GTE's cordless phones also fail to offer long-distance dialing compatibility—an interesting note, since GTE also owns and operates Sprint. In its Solitare series, Model 100 at $104.95 and Model 200 at $129.95 both have antennae built into the handsets and operate at an advertised range of three hundred and five hundred feet, respectively; both models have mute buttons, auto-redial buttons, and can be either desk or wall mounted. The more-expensive model offers a nine-number memory and intercom. GTE has also designed its familiar Flip-Phone to operate cordless: Model 300. It has the same features as the GTE Solitare 200 but provides a security feature in the base, an important function lacking in the lower-priced Solitare. The price is $199.95. The color of GTE's phones is almond. The warranty is for three months.

Sanyo has several models of cordless phones that offer tone dialing. Its Model TH 5400 lists for $179.95 and has a 256-digit coding security system, a nine-number memory, a thousand-foot range, an intercom, a mute, and auto-redial. The color is brown; the warranty is one year for parts and three months for labor.

Teleconcepts has one of the lowest-priced cordless phones on the market. Its Sentinel Deluxe lists for $89.95. The same phone operating on the newer FCC frequencies, its Sentinel Deluxe 46/49, is ten dollars more. Both phones have a seven-hundred-foot range, a ten-number memory, an intercom, auto-redial, and a security feature. The color is sand; the warranty is for three months.

TELEPHONES THAT DO TRICKS

The craze to add more gimmicks, geegaws, and electronic wonders to the basic telephone has led to a growing group of phones and communication devices that will drive a gadget-lover to his or her nearest phone store. Of course, these talking wonders don't come cheap. However, if money is no object, consider a few of the following:

The Desk Director System 600 may well be the most sophisticated communications system ever offered to the business person. For $995.00 The Sharper Image will send you an impressive black leather desk console, complete with phone, electronic dialer, a twelve-digit calculator, a clock, a calendar, and timing extras. A speaker phone, of course, means you can talk and shuffle papers at the same time. A hundred-number memory (twenty digits per number), an automatic redial that continuously dials a busy number until it's answered, a call timer, and the various more mundane features such as mute, volume, and last-number redial buttons, are standard. It has pulse or tone dialing, of course.

The Anova Communications Center does much the same as the Desk Director System 600 but is not part of a desk pad. It contains everything you need to make and receive calls and expands to become a total home control and protection system. Anova's built-in answering machine provides not one but seven outgoing message channels (including

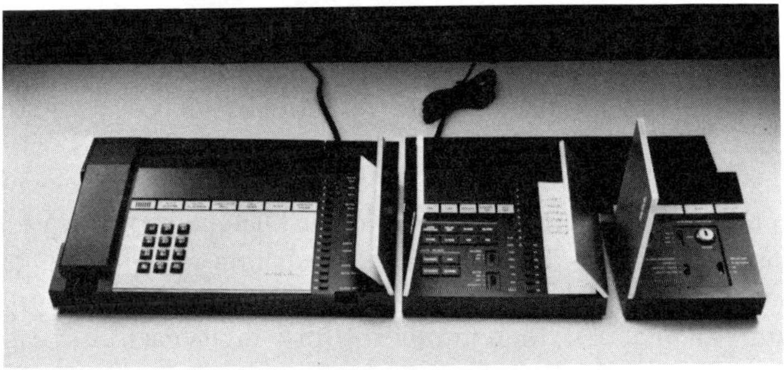

The Sharper Image

A Nova Communications Center

automatic tape-full message to ask callers to try again). The on-board microprocessor displays the number of calls received and the exact time and date of each message. You can receive messages when you're at home or away. The speaker phone is adjustable for single or group calls. If you want to record your calls, push a button. It stores and autodials sixteen numbers of fifteen digits each and will redial automatically if you reach a busy signal. It has a clock, a calendar, and a timer, of course, as well as pulse/tone dialing and all the other geegaws that you would expect on this advanced electronic marvel.

What separates the Anova system from other "simple" telephones is its ability to be united with separate Anova consoles. One unit, called the Control Center, gives you instant remote control and times automatic switching of every appliance in your home (up to 1600 watts). Another unit, the Protection Console, uses wireless detection sensors for intrusion, fire/smoke, and water. When tied to your Communications Center it will automatically dial the fire or police department, a doctor, or your neighbor. The security console has an independent key lock. The cost from The Sharper Image for the Communications Center is $399.00; for the Protection Console, $299.00; and for the Control Center, $199.00.

The Command Dialer, also available from The Sharper Image, gives you something the Anova system doesn't. You pick up the phone and say, "John." In seconds, you're conversing with John. "Doctor," you command, and your doctor's line is automatically dialed. Ah, the miracle of modern electronics! The Command Dialer's sophisticated circuitry breaks down human speech into a digital signal that can be read by the waveform analyzer. The memory stores a total of sixteen numbers (thirty digits each) and can recognize individual voices so more than one person can use the system. When you dial by voice, an LED flashes beside the name noted on the convenient log, confirming your command. Nonmemory push-button dialing is also part of the system as is the tone or pulse dialing plus the redial feature that redials a busy signal automatically fifteen times, pauses ten minutes, then repeats the cycle every ten minutes for an hour. Other features are a mute, an alarm clock with snooze bar, a ringer switch, and a battery light to let you know when your backup battery needs to be recharged. The cost is $249.00.

AT&T produces a sophisticated "build-it-yourself" telephone system. Buy a basic unit, then select from a combination of add-on modules and cartridges that are simply plugged into the basic unit. Ma Bell calls it the Genesis telesystem. The basic unit really can stand on its own. And it should, for the price. For $299.95 you get a desk console phone that includes a ten-number memory (sixteen digits each), a built-in clock and timer, an alphanumeric display that flashes information tied in with other phone functions, a speaker, plus an electronic notepad. The gimmick is the host of plug-in cartridges that you can buy for $39.95 each. For example, the electronic reminder cartridge reminds and remembers up to forty appointments, meetings, birthdays, and the like. The custom convenience cartridge makes call waiting and forwarding easier to use. The deluxe autodialer cartridge continuously redials busy or unanswered numbers automatically. You can buy each module sepa-

AT&T Consumer Products

rately or select a universal module for $199.95 that combines many of these functions.

AT&T also has an interesting assortment of phones "with a difference." Its Glow Phone, at $159.95, is a modern phone that looks like a square black box with a white base. Adjacent to the box is a phone cradle and a white telephone. Lift the receiver off its cradle, and presto—that translucent black surface lights up, and colored Op Art graphics suddenly appear, illustrating large letters and the number of the dial. Just touch a button on the surface of the box to dial a number. No fancy geegaws. Just a reliable telephone with Touch-Tone dialing.

Phone in a chest? Brookstone sells a high-quality, push-button touch-tone phone that is housed in a solid oak chest designed for an executive's desk. A brass plate on the hinged, wooden chest can be engraved with the owner's name. The price is $159.95.

For a phone in a child's room, how about Kermit the Frog? AT&T has Kermit sitting in his easy chair atop a pedestal with his feet resting comfortably on a modern footrest. The bright yellow phone is cradled on Kermit's outstretched foot. It has push buttons and Touch-Tone dialing. The cost is $99.95.

A flip-type phone that looks like a frog is also available from Adam York for $12.95. Pick up this frog and his eyes light up. No tricks; it has push-button pulse dialing. However, at that price you really can't expect Miss Piggy too.

A Coke bottle look-alike that is really a telephone is sold by Synchronics for $35.00. The push-button dialing system is hidden in the base; it works on both rotary and tone phones. Features include a last-number redial button, a mute button, and an extra-long cord.

Perhaps you would rather have a phone that looks like a duck and quacks rather than rings! The Decoy Phone looks just like a duck decoy, with a skillfully painted head and tail feathers. An incoming call alerts you with a soft quack. Pick up the plastic decoy's head and you're on the line. A four-foot cord automatically retracts when you replace the handset—or duckset, if you prefer. It has push buttons and true tone dialing, and is available from The Sharper Image for $249.00.

PHONES FOR EMERGENCIES

Although most phones are designed solely for two-way conversations, there are some interesting devices on the market that use your telephone lines to help you in an emergency. There are a host of burglar alarm devices hooked up to a prerecorded tape to call the police if a sensor is triggered. They are usually sold through specialized dealers who must either wire your home or tie electronic, sonic or infrared sensors to a master console that in turn triggers a tape to call for help. The Anova Protection Console, described earlier, is a system that can be set up by any handy person without hardwiring, a great deal of expertise, or hardware.

Security

One inexpensive remote device for bugging your home while you listen from anywhere in the country is The Listener, sold by Synchronics for $99.00. This sensitive instrument allows the user to call from any location and listen in on activities within the home or place of business. The microphone detects faint sounds, a turned-on radio, running water, or the presence of an intruder. Plug the device into a modular jack and activate the home bug with a pocket-sized security beeper.

Bugged? Have you ever had the feeling that someone was listening in on your conversations? If you believe that your calls are being tapped, or bugged, consider the Phone-Guard—a clever electronic device that flashes red when someone is listening in and will alert you instantly if there is a tap or bug on your phone. It also works if someone is listening on an extension phone or switchboard. It is installed by replacing your phone's existing mouthpiece with the Phone-Guard. It screws on or off simply and quickly and can be easily packed into a briefcase if you're concerned about being bugged in a motel or at the office. It's available from Synchronics for $49.00.

Another device to let you know if you're being bugged is the Bug Alert, sold by The Sharper Image for $29.00. It detects any eavesdropping on your line, whether from a lifted extension or a hardwired tap. When someone is bug-

THE MORE POPULAR CORDLESS TELEPHONES

1. Operating distance from base (in feet). 2. Pulse/tone switchable dialing. 3. Intercom. 4. Memory. 5. Ringer on/off. 6. Security feature. (Y=Yes; N=No)

Make and Model	*Price*	*1*	*2*	*3*	*4*	*5*	*6*
AT&T Nomad 2000	$ 99.95	50	Y	N	N	Y	N
AT&T Nomad 600	189.00	700	Y	Y	N	Y	Y
Cobra CP 450S	179.95	1000	Y	Y	Y	Y	Y
GTE Solitare 200	129.95	500	N	Y	N	Y	N
Mura MP–450N	99.95	700	N	Y	N	Y	N
Mura MP–680	169.95	700	Y	Y	Y	Y	N
Radio Shack ET–320	119.95	100	N	N	N	N	N
Radio Shack ET–330	179.95	500	N	Y	N	Y	N
Sanyo TH 5400	179.95	1000	Y	Y	Y	Y	Y
Teleconcepts Sentinel Deluxe	89.95	700	Y	Y	Y	Y	N
Uniden EX 2100	119.95	700	N	N	N	N	N
Uniden EX 4100	199.95	1000	N	Y	N	Y	Y
Uniden EX 4800	279.95	1000	Y	Y	Y	Y	Y

ging your line, a red light on the instrument goes out and the phone sound level is reduced. It plugs into any phone jack, is powered by the telephone current, and needs no batteries.

Is someone making unauthorized calls on your phone? A simple way to keep people from using your phone is to use a phone lock that fits over any desk push-button phone. It's available from Radio Shack (Model 43–116) for $7.95; it comes with two keys.

Use your phone to turn on lights or appliances remotely. Radio Shack sells a device it calls the Plug 'n' Power remote control system. A compact remote transmitter sends com-

THE MORE POPULAR CORDED TELEPHONES

1. Dial in handset. 2. Pulse/tone switchable dialing. 3. Ringer volume control. 4. On/off switch. 5. Memory. 6. Last-number redial. 7. Mute feature. 8. Hold. 9. Speaker. (Y=Yes; N=No; O=Available Option)

Make and Model	*Price*	*1*	*2*	*3*	*4*	*5*	*6*	*7*	*8*	*9*
AT&T Princess	$64.95	N	O	Y	N	N	N	N	N	N
AT&T Trimline	69.95	Y	O	Y	N	N	N	N	N	N
AT&T Touch-a-Matic 300/305	79.95	Y	Y	Y	Y	Y	Y	Y	N	N
Cobra ST–410	39.95	Y	Y	Y	Y	Y	Y	Y	N	N
GE Hotline	59.00	Y	N	Y	Y	Y	Y	Y	N	N
GTE Solitare IV	29.95	Y	N	Y	Y	N	N	N	N	N
GTE Solitare VI	54.95	Y	Y	Y	Y	Y	Y	N	N	N
Krone Compact 1000	130.00	Y	Y	Y	Y	N	N	Y	N	N
Mura CV 444	49.95	Y	Y	Y	Y	Y	Y	Y	N	N
Mura CV 464	69.95	Y	Y	Y	Y	Y	Y	Y	Y	N
Mura FP 200	79.95	N	Y	Y	Y	Y	Y	Y	N	Y
Panasonic KX–T2200	39.95	Y	N	Y	Y	N	Y	Y	N	N
Panasonic KX–T2340	89.95	N	Y	Y	Y	Y	Y	Y	N	Y
Radio Shack ET–270	39.95	Y	O	Y	N	N	O	Y	N	N
Teleconcepts Thinfone (desk model)	79.95	Y	Y	Y	Y	N	Y	Y	Y	N
Webcor 708SW	42.95	Y	Y	Y	Y	N	Y	Y	N	N
Webcor 757	79.95	N	Y	Y	Y	Y	Y	N	N	Y

mands over the phone to a telephone remote control center that is plugged into a standard modular telephone outlet and an AC line. The control unit operates up to eight lights and appliances by sending a signal over your house wiring to remote control modules. Just call your own number and send coded electronic signals to turn on or off lights, space heaters, air conditioners, a microwave oven, and the like. The remote control system costs $99.95. The modules are $16.99 each.

Fire, Flood, Heat, and Power Detection

If you're concerned about fire in an unattended home or office, AT&T has a valuable device called an Emergency Call System-Smoke/Fire Alert. It consists of a programmable console and either a transmitter that works with a previously installed smoke detector or a smoke detector transmitter. The console receives a signal from the smoke/fire detector and automatically dials the emergency phone number you have programmed into the machine. The console itself is only ten inches wide and five inches high. It can be desk or wall mounted and needs only a phone jack and an electric outlet. There are no wires between the console and the detectors. The system works with either rotary or tone phones. The cost for the console is $199.95; for the smoke detector transmitter, $59.95.

For a more complete emergency call system from AT&T, the Model 400 console, at $249.95, works with smoke/fire and medical transmitters to dial automatically two preprogrammed numbers for each emergency type. It uses voice synthesis to identify the emergency and alert the fire department, the police, a friend, or anyone. In addition, the Model 400 console can be hooked up to *environmental sensors*. Three types of environmental sensor are available at $29.95 each:

- *Temperature sensors* will call you or anyone you elect whenever the temperature in a room of your house goes below or above a predetermined setting.
- *Water sensors* can be located in a basement to alert you whenever you have a leak, break a pipe, or the like.
- *Power sensors* will tell you if you lose any 110V AC power.

These environmental sensors are a very good way to monitor a second home or keep track of potential problems in an office when you're away.

Medical Emergency

AT&T also has a medical alert system that allows an individual to push a button in a cigarette-pack-sized, battery-operated box that sends a message to a transmitter. The transmitter, in turn, automatically dials a doctor, a hospital, or whatever number you have preprogrammed, and announces the emergency. All AT&T emergency phones rely on voice synthesizers rather than taped messages, which can go awry. The console, which is about the same size as the smoke/fire console, costs $199.99; the small hand-carried transmitter, $49.95.

For shut-ins or persons with restricted movements the Body Guard, sold by The Sharper Image, also consists of a central transmitter that plugs into your phone jack. It has four separate channels marked: medical, police, fire, and general. You can preprogram a total of sixteen separate numbers, four per channel. A separate hand-held remote unit lets you call any number with the press of just one button. The pocket-sized remote will operate within three hundred feet of the base. The cost is $299.00.

Mura also has an emergency phone, Sage 1, Model SA–50, for $179.95. The emergency operation of the Sage phone is activated by a small, battery-operated transmitter. Pressing the transmitter's emergency button sends a coded signal to the telephone console that will then dial up to four different numbers, which you have programmed into its memory. It uses a computer voice synthesizer.

TELEPHONE ATTACHMENTS

In addition to the many types and varieties of telephones available, there are many attachments or add-ons to give you features that may not come with your present phones.

A Private Line

If you're interested in really protecting your privacy, PriveCode is an instrument that actually stops your phone from ringing and allows only selected callers to reach you. Whenever someone calls your number, before you hear a ring a voice asks the caller to enter his or her three-digit personal access code. The caller can either make the entry by pressing the numbers on a tone phone or, if using a rotary phone, by entering the code verbally, activating the machine's voice-responsive mode. The caller has three shots at getting the right code before being automatically disconnected. You can assign up to sixteen personal access codes. PriveCode has four different ringing patterns, each keyed to the first digit of the caller's code. Thus, you can program codes to signal which calls are for you, which for your spouse, and which for your children. And that's not all! A LED display shows you the caller's code so you know exactly who's calling before you answer. When you're out, the machine stores all the codes, so you know who has called; you can also hook it up to an answering machine if you want to receive messages. PriveCode has a phone jack for you to plug into your own phone (a telephone is not part of the package). It's available from The Sharper Image for $299.00.

For only $49.00, however, The Sharper Image has a

device it calls the Phone Censor. It is a small, white 8½-by-3-inch box that plugs into any modular phone jack. There are four separate codes that can be given to those you want to reach you. Each code causes the Phone Censor to ring with its own distinctive chime, letting you know before you pick up the phone who is calling. The device also lets you put a caller on hold from any phone in the house. And a built-in timer screens out all calls for one to ten hours on all your phones.

How many times have you had to get up and turn down the stereo when you answer the phone? Now an attachment called Phone Mute has been developed to relieve you of that burden. The gadget automatically lowers your stereo's volume whenever the phone rings or whenever you pick up the receiver, and it restores the volume when you complete your call. A master fade control allows you to reduce your stereo's volume by fifteen or thirty-five decibels. It plugs into an AC outlet, phone jack, and the stereo receiver. It's available from The Sharper Image for $59.00.

Ringer Silencer

Many of the accessories on the market simply allow you to add a feature that is standard with higher-priced equipment to a less-expensive phone. If, for example, your present phone lacks a ringer silencer, rather than run out and buy a new phone just to obtain this feature, stop in at your neighborhood Radio Shack and, for $6.95, buy a silencer (Model 43–125) that can easily be clipped to the back of any phone.

Outdoor Ringer

An outdoor ringer can be a useful accessory if you spend much time by the pool or in the yard. Radio Shack has a weather-resistant outdoor ringer that can be plugged into any modular outlet. With built-in volume control, Model 279–376 costs $21.95.

Multi-Line Controller

A multi-line controller that allows any corded or cordless telephone to accommodate two lines is sold by The Sharper Image for $39.00. This device, which requires two separate phone lines, has a red reminder light to let you interrupt one call to answer a second or put two calls on standby. Radio Shack has a device, Model 43–239, that performs the same basic functions for $29.95. AT&T also sells a two-line hold device that allows one phone to access two lines. In addition, AT&T's Two Line Hold 400, for $119.95, permits three-way conference calls using both lines.

— 7 —

Answering Machines

For many, an answering machine is not a gadget or a luxury but a necessary component to a telephone system. Some answering machines are available as independent units to be plugged into any modular jack; others are more complex, with telephones already attached and several tapes available for messages. They all have two separate tape sections: one to provide a message to callers, the other to receive a message when you can't answer the phone yourself. You record any message you want and, when you're out, turn on the machine and set it to "answer." The number of calls your machine can receive is limited by the length of the second "record" tape and the length of time you determine the tape should run before stopping. (This is a necessary feature to prevent someone from either speaking forever or getting cute with your time.) The machine also has a setting that allows you to decide how many times your phone will ring before it is automatically answered.

If you fail to pick up your phone before the answering machine is triggered, you and your caller will have to wait and listen while the machine delivers its pitch. It can be extremely frustrating to rush to the phone only to hear your own message blaring in your ear as you try to explain to the unknown caller that you are a real person and to be patient. Fortunately, there is a device called FoneAlone that solves the problem. It is a small electronic box about two inches square that plugs into your phone jack. Just pick up the receiver, and FoneAlone automatically silences any make or model of answering machine and resets it. You can install this gadget and never need to turn off your answering machine. It is available from The Sharper Image for $29.00.

More sophisticated answering machines have additional features that may be of value:

✆ *Voice-activated recorder switch, or VOX.* This keeps your recorder operating for the preset time as long as it "hears" someone talking. If your caller stops speaking for a few seconds, the tape automatically stops. This feature prevents you from having to listen to minutes of dead air or phone buzz when you play back recorded messages. After all, it only takes two seconds to say "Call your Mother!" Why waste another twenty or thirty seconds of tape after the message is delivered?

✆ *Remote Control.* You're away for the weekend and are expecting an important call. With remote control you simply call your own number and, either with the aid of a remote control device similar to the small box that controls your television or VCR remotely, or with a preset series of numbers delivered from any true tone phone, you activate your answering machine to play back the messages. Usually a remote feature allows you to advance or rewind the tape in case you missed a message.

✆ *Announce Only.* You're away for the weekend but want to leave your temporary number. No problem! Just leave a message that you can be reached at another number for anyone who calls your house.

✆ *Two-way Recording.* Although there are FCC regulations controlling the use and operation of telephone recorders, there are certainly plenty of them on the market if you want to bug your own calls. According to government regulations, "it is a crime under federal and state laws for any person to wiretap or otherwise intercept a telephone, unless that person has first obtained the consent of one or both parties actually participating in the call." So much for federal regulations. Some answering machines have the capability to record both ends of a phone conversation with or without consent. A few models, conscious of the law, beep every few seconds to let the caller know he or she is being bugged.

✆ *Memo.* This is an extension of the two-way recording mode: rather than tape a call, you can tape memos to yourself.

✆ *Message-received indicator.* These are on all answering machines. Some have a button that lights up when at least one call has been answered by your machine; others have counters that tell you exactly how many calls have been answered. Not all callers leave messages. There is something quite imposing about talking to an electronic device, and many people who are normally quite chatty suddenly freeze up when they realize that their priceless words are being recorded. By my personal estimate about a third of all callers simply hang up rather than talk to a mindless tape.

✆ *Battery backup.* Some of the answering machines come with built-in clocks and calendars. In case of a power outage the battery backup keeps both units running until power is returned.

All answering machines come with a microphone plus a message cassette to allow you to record your message. Message cassettes come in various types. The most common is the endless-loop, which plays your message for fifteen to thirty seconds, then automatically returns to the beginning to give the next caller the same words. Many machines allow you to decide how long your message will be. There is no need for the caller to hear a ten-second message and then wait another twenty seconds before he or she hears that familiar beep telling them that it is now their turn to talk to the machine. The lengthy delays undoubtedly account for many of those calls people never receive. On a few answering machines—the Code-a-Phone 1750, the Phone-Mate PM900, and the Sears 59140, for example—your message is permanently installed. The only way you can change it is to head for the nearest dealer or a local repair shop. You won't be reading about these machines in this book. A few machines on the market also come with a single cassette that serves as both a message-out and message-in recorder—another feature I would avoid when selecting your answering machine. If for any reason you want to substitute or replace either your message or your announcement cassette, Radio Shack sells 10-, 15-, 20-, 30-, 45-, and 60-second announcement cassettes for $5.95 each; their message cassettes are $2.39 for 60 minutes or $2.49 for 90 minutes.

How would you like someone to dial your number and have the phone answered by Cary Grant, Jack Nicholson, Ronald Reagan, or Richard Nixon? There is an interesting and fun series of prerecorded announcement tapes on the market produced by vocal impersonator Rich Little designed just for answering machines. He and other impersonators have teamed up to produce more than a dozen individual tapes, each containing a dozen twenty-second phony phone messages. (Write: Phonies, P.O. Box #2110, Cherry Hill, NJ 08003; or call: 609–424–6787.)

There's little doubt that more people are buying answering machines than ever before. An independent market study conducted in early 1984 showed that by the end of that year more than 3.4 million units would be sold, an increase of more than 70 percent over the previous year. The biggest sellers were Radio Shack, Phone-Mate, Panasonic, and Cobra. Let's take a look at these and a few other models now on the market.

Radio Shack offers a choice of five answerers. Its least

expensive, the DuoFone TAD–111A–43–246–1, is a solid, basic unit. It records up to 120 incoming messages and has an announcement mode, ring delay, and built-in microphone. The cost is $119.95. Radio Shack's newest model, the DuoFone TAD–215–43–319, offers you a great deal more. It should; it lists for $239.95. A built-in voice synthesization lets the answerer communicate with you directly. Just dial your number from any phone, then verbally identify a two-digit code number that prevents unauthorized access to your machine. You will be asked a series of questions: "Play back new messages?" "Play back all messages?" and many more. You respond by saying "yes" or remaining silent. It also has "announce only," two-way recording, a memo mode, a phone jack, and a digital message-receive indicator. Radio Shack has a more expensive model, the DuoFone TAD–150–43–248, that lists for $299.95 and gives you a fluorescent display showing the message number plus the time and date. It also has a four-digit security code. Dollar for dollar, the TAD–215 is probably a better buy.

Phone-Mate has been a leader in the field for years and has pioneered several types of machines. In 1983 it introduced a completely new line, the IQ series, with new additions to the line in 1984 and 1985. The IQ 935, at $109.95, is a voice-activated, dual-cassette system that replaces the Model PM900 mentioned above. Unlike its predecessor, you can record and personalize your own messages. The voice-activation feature eliminates long pauses between messages. The long-life message tape allows callers to leave a message as long as five minutes—shorter if you so program the machine—and cuts off automatically when the caller hangs up. A last-message indicator tells you when you've played back the last call. It has no remote; if you want a remote from Phone-Mate, the cheapest is IQ 940, at $159.95. It allows you to retrieve messages from your machine from any true tone phone. If you don't have a tone phone, however, you'll have to wait until you get home. If

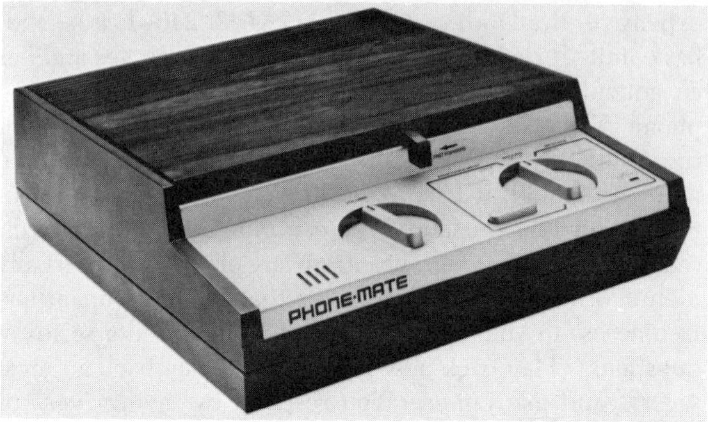

Phone-Mate, Inc.

for some reason you want to change your announcement tape, this can also be done from a tone phone. The IQ 2845, for $189.95, gives you the same features as the lesser-priced models plus a remote backspace feature so you can replay messages you didn't quite catch the first time. You also get a digital-display message counter.

Panasonic has a well-designed series of answering machines that provide the same basic features as their competition. Its least-expensive model, KX–T1410D, lists for $99.95 and is completely voice activated. Both incoming calls and your announcement message can be timed. Rather than a digital-display message counter, this model uses a LED light that blinks up to 15 times to show the number of messages received. It has no remote playback. Model KX–T1425, at $179.95, however, does provide a remote playback using any true tone telephone. It also allows you to reset the machine, repeat individual messages, and leave a personal marker message. As an added feature, according to Panasonic, you can "use the answerer to listen for intruders when not at home by using the 'room-monitor.'" This gimmick hooks up the VOX, or voice activated recorder, to the microphone used to record your memo or announcement messages. When the VOX picks up sounds in the room, it records them. The company, however, doesn't say what you're supposed to do with a tape recording of burglars trashing your house, unless, of course, they're considerate enough to yell out their names and addresses while in the act of housebreaking. Also designed for beeperless remote-control operation is Panasonic's Model KX–T1730, at $279.95. This unit also has a remote feature for use with tone phones, as well as a synthesized voice confirmation announcing calls that have been recorded. The mechanical voice precedes each message, noting the time of the call. Both models incorporate a "ring control" feature to beat Ma Bell out of a toll call when calling your machine long distance for messages. The machine will answer immediately if you have any messages awaiting. If not, it will wait three rings before making the two-way phone connection. For a complete Panasonic telephone system, including a phone, an auto dialer, and a beeperless answering device with a speaker phone attachment, take a look at top-of-the-line Model KX–T2425, for $269.95

Sanyo has a reasonably priced series of sleekly designed,modern-looking answering machines. Model TAS 10, at $99.95, will record up to thirty thirty-second messages; however, it lacks a remote feature. Sanyo's Model TAS 30, at $169.95, does come with a beeper-remote and is voice activated to record up to four minutes of incoming messages. TAS 4000, at $219.95, automatically resets itself after you call in for messages and gives both the time and the date of message receipt.

Cobra has a series of low-priced answering machines all selling for under two hundred dollars. Its least-expensive model, AN–3100, at $99.00, lacks a remote feature. However, Model ANP–3770, at $179.95, is a full-feature, fully automatic microprocessor-control machine including a remote key and a toll-saver feature similar to Panasonic's.

Cobra, incidentally, maintains a company hotline for consumers who have questions or problems with any of their products. The number is 800–COBRA–22.

GTE offers two answering systems. Its Model 3100, at $129.95, is a dual-cassette unit without a remote feature. Model 5100, however, at $179.95, has a keyless remote plus a two-way recording feature.

Webcor has a relatively inexpensive feature-packed answering machine: Model 1010, at $199.95. It has a beeperless remote that controls all functions by audibly responding to a series of beeps. A computer determines whether the individual calling in for messages is authorized to receive them. It also has a digital clock, a stopwatch, the VOX feature, and a message counter.

AT&T must not be left out. With the breakup of Ma Bell, AT&T's Consumer Products Division began producing equipment never before available from the phone company. Their answering machines are part of the new product line. AT&T's least-expensive unit, Model 2000, at $89.95, does not have a remote playback feature. For $159.95, however, Model 2300 comes with a remote key, the ability to record both sides of a phone conversation, and a message counter. The top of the line is Model 5500, at $279.95, which was introduced at the end of 1984. This answering machine combines a telephone with a dual-cassette, remote access device. The features include a twelve-number memory, last-number redial, pulse/tone switchable dialing, VOX, and the option to have messages call-forwarded to another number.

THE MORE POPULAR TELEPHONE ANSWERING MACHINES

1. Call Screening. 2. Voice-Activated Recording. 3. Remote Playback. 4. Digital-Display Message Counter. 5. Toll-Saver. 6. Two-Way Recording. (Y = Yes; N = No)

Make	*Model*	*Price*	*1*	*2*	*3*	*4*	*5*	*6*
Radio Shack	TAD–111A	$119.95	Y	Y	N	N	N	N
	TAD–215	239.95	Y	Y	Y	Y	Y	Y
	TAD–150	299.95	Y	Y	Y	Y	Y	Y
Panasonic	KX–T1410D	$ 99.95	Y	Y	N	N	N	N
	KX–T1425	179.97	Y	Y	Y	Y	N	Y
	KX–T2425	269.95	Y	Y	Y	Y	Y	Y
Phone-Mate	IQ 935	$109.95	Y	Y	N	N	N	N
Sanyo	TAS 10	$ 99.95	Y	N	N	N	N	N
	TAS 30	169.95	Y	Y	Y	Y	N	Y
Cobra	AN–3150	$ 99.00	Y	N	N	N	N	N
	ANP–3770	179.95	Y	Y	Y	Y	Y	Y
GTE	3100	$129.95	Y	N	N	N	N	N
	5100	179.95	Y	N	Y	Y	N	Y
Webcor	1010	$199.95	Y	Y	Y	Y	Y	Y
AT&T	2000	$ 89.95	Y	N	N	Y	N	N
	2300	159.95	Y	Y	Y	Y	N	Y

— 8 —

Automatic Dialers

Few people remember those old days when, to make a call, all you had to do was pick up the receiver and, when the operator said, "Number please?" pass on no more than a one-word exchange plus four digits. Those days are long gone, along with many of the romantic exchanges—Trafalgar, Murray Hill, and Beekman. Now if you want the personal touch of an operator's voice, you have to pay a great deal extra for it. To make a simple long-distance station-to-station call may now mean dialing as many as twenty-five digits. Speak to a live operator and request a long-distance, or a local, telephone number, and you are charged for the visit. Now, however, there is a convenient and fairly inexpensive way to dial a neighbor next door, a relative across the country, or a business associate in Japan. The direct or automatic dialer gets you your number with the press of just one, or at most two, buttons.

The automatic dialer is an extremely simple device to use. This small box lists the names of anywhere from a dozen to several hundred individuals or companies, with a button adjacent to each name. The box plugs into any modular outlet; your phone is then plugged directly into the dialer. To make a call, just press a button and the dialer does the rest: it waits for a dial tone, it provides the necessary sequence of tones to reach your number, and, if you're using a long-distance service such as Sprint, MCI, or SBS Skyline, it waits until it reaches your local access number before completing the remainder of the numbers. A "pause" feature, which allows the local access number to be reached before continuing the dial sequence, is not available on some inexpensive automatic dialers. But it is a feature worth holding out for. Many of the more complex telephone systems described in earlier chapters have dialers as part of the total phone package. AT&T's Genesis and Touch-a-Matic phones, for example, will each hold up to thirty-six preprogrammed numbers available at the touch of a button.

Radio Shack produces an inexpensive, reliable dialer for $49.95, the DuoFone–100. (This model is often on sale for under $35.00.) It stores up to sixteen numbers of fifteen digits each—which means that to use Sprint or MCI you must press one button to reach the local access number and a second to complete the tag-end of the call. The dialer doesn't have a pause button; however, the DuoFone–232, at $99.95, does. This more expensive model also gives you a multi-function LED display that shows the number dialed, the time of day, and, if you want to time your call, a stopwatch. Both dialers work with either pulse or tone dialing. For $119.95, Radio Shack has an automatic dialer that is hooked up to a modular outlet or to the incoming telephone line. When connected to the incoming line at the standard network interface (SNI), the unit acts as a memory box for all the phones on that line. Just dial a preprogrammed name or number and the "box" does the rest. It will also redial busy numbers every ten minutes for up to two hours. Unlike the Deamon Dialer, this device must be used with tone phones only.

The Deamon Dialer is one of the most ingenious gadgets on the market. It is manufactured by a small Boston-based company called Zoom Telephonics. The device is a small black box that is easily installed to the incoming telephone line at the SNI, the point where the phone company's wires enter your house or apartment. (The lower-priced unit, 93H, is plugged into any modular outlet if you want to limit the Dialer to single-phone use.) Detailed and easy-to-understand instructions come with the units. Model 176T comes with a series jack in case your current installation requires it. A second connection is required to any electrical outlet. When you attach the Deamon Dialer to the SNI, any phone in the house automatically can use the services of the Dialer.

When hooked into the Deamon Dialer, a phone, whether pulse or tone, can then:

- Call any number in the world by simply pressing, or dialing, your party's initials or any coded number you elect. Long-distance services' numbers (of MCI, Sprint, and the others) can easily be programmed into the unit. You need press only a single coded number, not a separate number for MCI and a second number for your party. The Dialer will dial your local access number, wait for the computer tone, and then complete the numerical sequence.
- Reach a busy signal? Press "2" and your Deamon Dialer will dial that number for up to two hours. When the line becomes free, you will hear a beep through your phone. Pick it up and speak with your party. If someone tries to reach you while your Deamon is dialing, your phone will ring normally and the Deamon will be automatically disconnected.
- No one answers? Press "4" and the Deamon Dialer will call every ten minutes for up to ten hours and let you know when your party finally answers.
- Last-number redial? Just press "1."

The Deamon Dialer is available in two models. The 93H can dial up to ninety-three seven-digit numbers or fifty-five twenty-three-digit numbers. The cost is $130.00.

Model 176T has a series jack that, when connected to the SNI, allows one Deamon Dialer to be used by all the phones on a line. The cost is $200.00.

AT&T has a sixteen-number (sixteen digits each) dialer for $64.95 (TeleHelper 1600) and a sixty-four-number (sixteen digits each) dialer/speaker phone (TeleHelper 6400) for $139.95. The latter unit has a vacuum fluorescent display screen that shows the time, the day, the date, the number dialed, and the time elapsed on the call. Both of these dialers require punching two buttons for long-lines service: one to reach the service's access number, the second to dial your personal code, area code, and telephone number.

Panasonic has several efficient dialers, all of which have a capacity of but sixteen digits per number. Model KX–T1225D, at $139.95, stores thirty-two numbers; Model KX–T1235D, at $159.95, stores sixty numbers. Both are compatible with tone phones and both have a redial button and a monitor speaker. A pause feature and the ability to combine numbers for use with long-lines services means you can still dial a thirty-two-digit number with the push of one button.

PORTABLE DIALERS

There are several interesting devices on the market that either allow you to use a portable tone dialer with a rotary phone or have the capacity to store numbers within a pocket-size auto-dialer. For people who currently have a rotary phone and want the advantages of the long-lines services without buying new tone phones, a tone dialer is an answer. A small battery-operated push-button dial pad is placed on the speaker end of any telephone—tone or rotary—and, when you punch out your desired number, a tone is generated. Radio Shack sells a 4-by-2⅜-by-1⅛-inch pocket tone dialer for $24.95.

Ma Bell has a pair of small take-along dialers that supply those musical tones that you need for the long-lines services or, in the case of AT&T, provide a phone company credit card number via tone. The basic pocket dialer, Model 100, has no storage capacity for memory—just a basic dial pad. The cost is $24.95. If you want the memory feature, AT&T's Model 110, at $39.95, gives you a choice of up to nine different numbers (sixteen digits per number), plus a redial feature.

Panasonic has a similar size tone dialer, Model KT–A70, for $49.95, that doubles as a remote pocket dialer for any Panasonic beeperless answering machine. Dial the answering machine from any phone, and when the machine picks up, the dialer is held to the mouthpiece to transmit a preprogrammed security tone code that prompts the answering machine to play back the messages. The dialer also stores up to thirty telephone numbers of thirty digits each.

SOME POPULAR AUTOMATIC DIALERS

1. Number Storage (phone numbers/digits per number). 2. Pause Button. 3. LED Display. 4. Redial Button. (Y = Yes; N = No)

Make	*Model*	*Retail Price*	*1*	*2*	*3*	*4*
Radio Shack	DuoFone–100	$ 49.95	16/15	N	Y	N
	DuoFone–232	99.95	32/15	Y	Y	Y
AT&T	TeleHelper 1600	$ 64.95	16/16	Y	N	N
	TeleHelper 6400	139.95	64/16	Y	Y	Y
Panasonic	KX–T1225D	$139.95	32/16	Y	N	Y
	KX–T1235D	159.95	60/16	Y	N	Y
Zoom Telephonics	Deamon Dialer 93H	$130.00	55/23	Y	N	Y

In addition to activating beeperless remote answering machines from non-tone telephone, it may also be used to access long-distance discount services such as Sprint and MCI, or it may simply serve as a small portable auto-dialer.

For something a bit fancier the Dial-It, manufactured by Dictograph and available from The Sharper Image for $69.00, combines a small auto-dialer with a watch and calculator. The tiny device measures 3 by 2¼ by ½ inches and weighs only four ounces. It stores and dials one hundred numbers of thirty-two digits each. It has one-button redial of busy numbers and an electronic security lock, plus you get a quartz-accurate alarm clock, a stopwatch, and a full-function eight-digit calculator with a memory. Numbers are displayed on a LED screen. If you want to program numbers other than phone numbers, such as Social Security or identification numbers, here's a good way to store and carry them. This clever electronic machine won a design award at the 1983 Consumer Electronics Exhibition.

If remembering coded numbers is too difficult, The Sharper Image also has a device called the Ultra-Dial, which is a small remote dialer that works by keying names, not numbers. The 8½-ounce shirt-pocket-size dialer stores eighty-five names (up to eight letters long) and instantly dials the accompanying number (up to thirty-two digits) with a single touch. You can also program the necessary pauses for the long-lines services. It also remembers credit card numbers, Social Security numbers, and the like. It has last-number redial and automatic shut-off. It works with true tone phones, and costs $79.00.

If you don't care about the portability feature but still want to have a tone adapter for your rotary phone, consider SoftTouch, sold by Synchronics. This is a tone auto-dialer in the shape of your phone's circular mouthpiece. Screw off the speaker, or mouthpiece ring, and screw on the SoftTouch auto-dialer. All your programming and dialing is done right from the handset. Once you enter digits into the solid-state memory it stores and dials them with the press of a button. It comes in three models: SoftTouch 3, at $39.00, has three memories with one-button auto-dialing and a redial button; SoftTouch 23, at $59.00, adds twenty regular memory positions of eight digits each, stackable to 160 digits; and SoftTouch 83, at $79.00, has eighty regular memory positions of eight digits each, stackable to 640 digits.

— 9 —

Intercoms, Pagers, Amplifiers, and Telephone Recorders

INTERCOMS

How many times have you been in one part of the house and had to reach someone in another? A phone call, a problem with the furnace, someone at the door—it happens all the time. You have several choices. You can yell, which is what most of us do, run upstairs looking, or just assume that he or she is not home and try to handle the situation as best as possible. There is another option—an intercom.

If you want to get fancy, there are professional communications companies that will come to your house, string new wiring through walls and ceilings, and leave you with an elaborate series of boxes containing speakers, buttons, and microphones as well as a hefty bill. Far less expensive are the intercoms that use existing electrical house wiring. However, there are other ways to reach John or Judy in that upstairs bedroom when you're busy elsewhere.

Perhaps the simplest and least-expensive intercom system is your cordless telephone. Many cordless phones have an intercom feature that can often be worth the price of the phone itself; it's certainly cheaper than installing a new wired home system. A major advantage of the cordless phone intercom is its portability. You can't tote a wired box around the house or into the yard. To reach Judy by the pool or next door with a cordless phone, just press the "page" button on the base unit, and Judy's phone buzzes. It couldn't be easier. If Judy wants to contact you, she also has a "page" button on her portable phone.

The cordless telephone intercom is controlled at the base. A button switches the system between telephone and intercom. If the base is turned to "intercom," the cordless phone cannot be used as a telephone; it can only communicate with the base unit. Incoming calls cannot be answered either at the base or at the cordless unit until the base switch is turned to "telephone."

If you don't want to invest in a cordless phone, consider the intercom systems that work off your house wiring. For example, Radio Shack sells a three-station, two-channel (so you can speak to one station without being heard by the other) system for $129.95—Model 43–214. Each station is plugged into an AC outlet in the rooms you want your intercoms. An automatic squelch feature keeps stations silent until called. Each station has call-tone and talk controls, a channel selector, plus LED "talk" and "receive" indicators. Radio Shack also has a two-station wireless intercom, Model 43–212, for $69.95. Each station has a talk bar, call-tone, and volume control.

If you want to give your present corded phones the ability to double as an intercom as well, consider the TalkTo system sold by The Sharper Image. This intercom installs in just a few minutes by plugging a compact page unit into each of your present telephone outlet plugs. With this system you can put a caller on hold, dial a single-digit code, and reach any phone in the house. You can also make house-wide announcements over the page unit's built-in microphone and loudspeaker. The sensitive microphone lets people answer from across a room. Each page unit also gives you one-button dialing of nine emergency numbers. A privacy circuit prevents other extensions from listening to your calls; for conference calls, however, all phones can be linked by the flick of a switch. In rooms without telephones a separate monitor unit can be used for an intercom, paging, and continuous monitoring of room sounds. If anyone really likes Muzak, a radio or tape can be plugged into the system to spread it throughout the house. This is a far more sophisticated electronic device than a simple intercom, as the price reflects: the TalkTo system, including a master control unit, one page unit, and one monitor unit, costs $399.00; additional page units or monitor units sell for $89.00 each.

A simple buzzer system can also be installed throughout a house to tell a family member that a call awaits. This solution, however, does require hardwiring and is often more trouble than it's worth. An inexpensive intercom using AC wires may be the best solution.

PAGERS

Although pagers are not part of your telephone system, they should be briefly mentioned. A paging device that is not connected to your phone, as described in Chapter 6, is usually part of a more involved service sold by a professional paging company; look up "Paging and Signaling" in your Yellow Pages. If you subscribe to a service, you can buy or lease one or more small beepers and pay a monthly subscription fee plus a charge for every page received. Each beeper has its own unique radio frequency. Either a central office transmits your unique signal whenever there is a message for you, or simply dialing a special number or code assigned to you will trigger the system to page your beeper. To prevent a loud beep during a church service or while at the opera, a switch silences your beeper until you are ready to receive messages. Some of the more sophisticated beepers do more than beep. They have small LED displays to deliver the message directly: "Call home," "Call 212–555–1212," "Meet Jane at the Plaza at five." Costs for message displays range from $300 to $450 plus a per-message charge; rental fees vary from $35 to $40 per month. In 1984 the Federal Communications Commission announced that it would assign three national frequencies to a few applicants, thus permittng a coast-to-coast paging network enabling your beeper to function anywhere in the country. Prior to national transmission, which is expected to be in operation by the end of 1985, beepers only operate within a limited radius.

In the future, paging services hope to use satellite transmission to bounce call-signals around the world. In some parts of the country Radio Shack sells a small pocket pager for $99.95 plus a monthly and per-call charge from its local paging service.

AMPLIFIERS

Telephones designed for the hearing impaired are, unfortunately, few and far between. Most phones on the market have a control to regulate the volume of the ring, but very few allow you to control the level or quality of the incoming voice. AT&T, however, does sell a volume control handset that is compatible with any modular telephone. A control wheel in the handle amplifies sound to the desired level. The cost is $29.95.

To amplify the sound in an existing telephone, Radio Shack has a handset amplifier, Model 43–237, for $19.95, that attaches to most phones and operates with a nine-volt battery. Simply plug the unit into the handset and incoming signals are boosted. Another unit sold by Radio Shack, Model 43–238, is an earpiece amplifier. Also battery operated, this device ties around the existing earpiece and serves as a small voice receiver and amplifier. It has a volume control and, according to the manufacturer, "boosts caller's voice up to five times." It is easily removable and can be stuffed into a bag or pocket for use on other phones. The cost, with battery, is $11.95. Another solution to the inaudible sound problem is the use of either a speaker phone or a separate speaker attached to your phone (see pages 87–90). The problem with speaker phones, of course, is the quality of sound. For the hearing impaired it doesn't help matters to amplify the sound only to have it garbled. There is also a loss-of-privacy factor that has to be considered when you use a speaker phone. It's all right in an office, but everyone doesn't want their phone conversations broadcast about the room.

TELEPHONE RECORDERS

The simplest way to record your telephone conversations is with the recording feature on many telephone answering machines. However, there are alternatives. Although the question of recording, or bugging, a conversation gets into a quasi-legal area, as mentioned on page 113, there are devices on the market that enable you to record your conversation, or for that matter any two-way telephone conversation, on your line without either party knowing about it. Obviously, bugging a phone, even if it's your own line, without both parties being aware of your activities, is a no-no. Caveat emptor!

If, for example, you own any type of tape recorder—stationary or portable—you can buy a multi-extension recording control device at Radio Shack, Model 43–236, for $24.95. The unit is plugged into any modular telephone outlet and is connected to your tape recorder by a pair of plugs. One plug is inserted in the microphone input, the other into the remote input. That's all there is to it. A switch on the unit lets you turn the recording on and off and control "record" and "playback." Put a cassette tape in your recorder, switch the recorder on, and wait. Whether you're in the house or not, every time a telephone is picked up on your line the recorder automatically turns on and records. Hang up, and the recorder shuts off. If you only want to record from a single phone, Radio Shack has a less-expensive model (43–228), at $19.95, that works the same way.

—10—

Cellular Telephones

The newest development in telephone technology is the mobile, or cellular, telephone. Now in most cities you can be within an arm's length of your telephone twenty-four hours a day—at home and on the road. For salespeople and business people who spend hours a day traveling in their cars, the coming of the cellular phone has changed business practices. Instead of roaming the radio dial, they can conduct business on their car telephones. Whether a phone in the car is a boon or a potential headache, only time will tell. Just as you can call a client, unwanted calls can now reach you on the highway without the safeguard of a spouse or secretary to act as a screen.

Although the big rush to install phones in cars has been a recent trend, putting radio telephones in vehicles is far from a new development. In 1921 the Detroit police department tracked crime by car phones and in 1946 the Bell system introduced in St. Louis what was called a "public correspondence system," designed to permit communications among a variety of users via radio. The FCC allocated six channels for the system in the 150-MHz range. A problem that existed then, and to some extent still is very much as yet unsolved, is the number of people, businesses, and federal agencies that want to use a finite number of radio frequencies. Obviously, with a limited number of wavelengths, or frequencies, available, not every individual or company will receive the FCC's permission to transmit. Thus, owning a radio telephone has been an expensive and, up to now, hard-to-obtain luxury. Because of the limited number of frequencies, the FCC has assigned only forty-four mobile telephone channels, all transmitted from a single antenna. One channel is necessary for transmitting; a separate channel is needed for receiving. Ironically, Ma Bell called the system IMTS, for Impoved Mobile Telephone Service. You can only imagine what the service was like before the "improvement." With IMTS only two dozen callers could use their telephones at one time. Anyone with more than one teenager knows what getting a dial tone must have been like. In New York City, for example, there are only two thousand individuals who are licensed to use the IMTS service. Pick up your car phone in the Big Apple, and you usually have to wait thirty minutes or more before there is a free line. For more fun and games consider that to acquire a permit to wait a half hour with a phone held to your ear may take two or more years. To complicate matters further, with but one transmitter covering an entire metropolitan area there are serious transmission problems. If a tall building gets between your car and the transmitter, you can easily get an earful of static. Signals lost entirely are not uncommon with IMTS. No wonder cellular technology has had such a huge impact.

WHAT IS A CELLULAR PHONE?

Instead of a central transmitter and a dozen frequencies to blanket an entire city, the cellular concept divides a metropolitan area into many small honeycomb-like geographical units called "cells." Each cell is serviced by its own transmitter and antenna, with its own set of allocated frequencies allowing reuse of each channel for different conversations many times. The cellular approach eliminates the need for high-powered radio transmission by carrying the conversation over regular telephone lines to a cell site near the mobile customer. As a vehicle moves from cell to cell within the city, the call is "handed off" to an adjacent cell. This computer-controlled system monitors every call and maintains vocal quality without any interruption. Neither caller nor receiver has any idea when the car crosses from one cell to another. The sound quality is comparable to speaking over any standard desk phone. To alleviate the problem of limited frequencies, in March 1982 the FCC increased the number of mobile channels to 666, opening the gates to increased service. Instead of only a few thousand parties using a mobile phone, cellular technology opens service to tens of thousands. In Chicago, for example, the cellular system handles up to fifty thousand calls per hour, compared with IMTS's maximum of fourteen hundred.

This new technology is a potential windfall not only for those who want a car telephone but for dozens of companies as well. Between 1983 and early 1985 the FCC has received more than fifteen hundred applications for licenses in the country's top markets. Rather than create a panic to get in on licenses, the FCC ruled that for every metropolitan area only two licenses will be granted: to a public company equipped to provide cellular technology and to the local wire line telephone company. The major companies competing for local licenses include MCI Communications, Metromedia, and Western Union.

COSTS AND EQUIPMENT

Although cellular telephones are a great deal cheaper than those that are limited to IMTS, they are still considerably more expensive than your home or office desk model. Aside from the cost of buying and installing the new phone, service charges eat a healthy hole in any budget. For example, in Chicago, Ameritech Mobile Communications charges a monthly service fee of $50.00 plus a per-minute charge of $0.40 for calls placed between 7:00 A.M. and 7:00 P.M., or $0.24 for calls made between 7:00 P.M. and 7:00 A.M. In the Washington, D.C., area, however, the Bell Atlantic Mobile System only charges a $25.00-per-month service charge. What is happening, obviously, is that, with two competing services in every metropolitan area going after a million-dollar pie, prices have to come down.

In addition to the monthly and per-call charges you also have to acquire a telephone plus the ancillary equipment needed to operate on the road. There are four basic components to cellular hardware: the telephone instrument; a power source—usually a five-pound heavy-duty battery—mounted in the trunk; a cable connecting the power source to the telephone; and a transmitting/receiving antenna.

The major manufacturers of cellular phones include Motorola, Panasonic, AT&T, and Western Union; Mura announced its introduction into the cellular market in the summer of 1984. Each major manufacturer of cellular phones is trying its best to one-up the guy next door. One of the largest manufacturers, Motorola, for example, became the first to announce a "portable" cellular phone. Its Model DYNA T–A–C 8000X incorporates the cordless concept in a cellular phone. The handset contains a rechargeable battery that can hold up to a one-hour charge. Put the handset in your briefcase and amaze your friends as you receive a phone call in the middle of a business lunch.

Making a call from a car is basically the same as calling from your bedside. There is one major difference: As you punch the number you want to call, a tone signal is not immediately sent over the car's radio telephone. Rather, the number is held in the phone until you press a "send" button. The number is usually displayed on an LED atop the telephone. Make a mistake in dialing—press the "clear" button, then dial again. To hang up you press the "end"

The dial-in-handset control unit of the AT&T Cellular Telephone System 1000 is compact and has built-in features that include automatic dialing of stored numbers and automatic recall, both activated at the touch of a button. The cellular phone is shown installed in a Buick Riviera.

AT&T Consumer Products

button. It's really quite easy. All cellular telephones are equipped to handle tone calls and have a built-in memory for up to ten numbers. To dial a stored number, press a button.

To prevent unauthorized use, all cellular phones are given a unique code number. This number is stored in the phone company's computers. If either your car or phone is stolen, report the fact to the phone company, and the computer picks up and prevents any further calls to or from the marked phone.

There is a problem with cellular technology, however. At present, there is no standardized method of transferring calls between different metropolitan areas. Since each area is operated by a separate company with its own computers, calls are difficult to switch from city to city. However, with advance notice it may be possible for another area to input your phone's codes, thus enabling you to operate within a zone beyond the range of your home base. Drive from Boston to Washington, and you cross through zones covered by three different telephone companies. If you're a mobile subscriber in Boston, your phone won't receive calls too far beyond the Charles River unless you make separate arrangements with carriers in each city you plan to pass through. In some areas you can subscribe to a "roam" feature, which automatically hooks you into the computers controlling several cities. Chicago and Milwaukee, for example, are both served by Ameritech Mobile Communications. With several cities served by the same cellular carrier, it is much easier to switch calls. The same is true among Baltimore, Washington, and Philadelphia, which are all served by the Bell Atlantic Mobile System. New York and Philadelphia are only a hundred miles apart, yet they are served by two different companies.

There is a further catch. FCC requirements state that a mobile phone owner *must* be free to transmit anywhere. This means that you can drive from New York to Washington and "probably" make outgoing calls in each city that has a cellular network. The catch is that only your home network is computer equipped to monitor the number of calls you make and bill you accordingly. Be assured that this lack of standardization will not last long. It is but a matter of time before your cellular phone will be able both to transmit and to receive calls wherever there is a cellular network.

Buying and installing cellular equipment is much like buying any telephone except that the price is ten to twenty times greater. Remember that you buy only hardware. Getting hooked into the phone lines means a separate arrangement with one of the two carriers in your city. Some cellular phone dealers secure blocks of local telephone numbers and will gladly assign you a number and assist you in hooking up with a phone company. Prices of hardware vary although there are fewer options with a mobile phone than with your house phone. The price of a mobile phone as of September 1984 ranged from two thousand to three thousand dollars, depending on the manufacturer. In addition, plan on spending about $250 for installation. Some installers include an antenna; others expect you to pay another fifty to one hundred dollars for this necessity.

According to AT&T, the current price structure for cellular phones is much like that of calculators when they first hit the market. Expect cellular phone prices to drop at a rate of about 10 percent a year until they level off at the one-thousand-dollar mark. When shopping for a cellular phone, keep three questions in mind:

- Who gives you a phone number?
- Who installs your complete cellular package?
- Who will service your phone?

Many of the manufacturers vying for your business will promise all three. Be sure they can deliver. Both AT&T and Western Union, for example, assure all customers that service is available in every city now handling cellular. They also will install your phone and take care of your phone number.

Keep in mind that every city must have two companies providing communications services. Although the FCC has not licensed the second carrier—in addition to the present line company—in every one of America's metropolitan areas, there is no reason why you can't start shopping now. Contact your phone company and get its rates. It won't be long before a competing service is in the picture offering you a "better deal."

Companies serving many of the nation's metropolitan areas include:

Atlanta	BellSouth Mobilnet. 404–956–2400
Baltimore	Bell Atlantic Mobile System. 800–255–2355
Boston	Nynex Mobile Communications. 800–431–2300
Buffalo	Nynex Mobile Communications. 800–431–2300
Chicago	Ameritech Mobile Communications. 800–662–4531
Cincinnati	Ameritech Mobile Communications. 800–622–4531
Cleveland	GTE Mobilnet. 800–828–7280
Dallas	Southwestern Bell Mobile System. 800–331–0500
Denver	Newvector Communications. 800–626–6611
Houston	GTE Mobilnet. 713–821–8900
Indianapolis	GTE Mobilnet. 713–821–8900
Kansas City	Southwestern Bell Mobile System. 800–331–0500
Los Angeles	PacTel Mobile Access. 800–321–4853
Miami	Bell South Mobility. 800–351–2400
Minneapolis	Newvector Communications. 800–626–6611
Milwaukee	Ameritech Mobile Communications. 800–622–4531
New Orleans	BellSouth Mobility. 800–351–2400
New York	Nynex Mobile Communications. 800–431–2300

Pittsburgh	Bell Atlantic Mobile System. 800–255–2355
Philadelphia	Bell Atlantic Mobile System. 800–255–2355
Phoenix	Newvector Communications. 800–626–6611
Portland	GTE Mobilnet. 800–828–7280
San Diego	PacTel Mobil Access. 800–626–6611
San Francisco	GTE Mobilnet. 800–828–7280
San Jose	GTE Mobilnet. 800–828–7280
Seattle	Newvector Communications. 800–626–6611
St. Louis	Southwestern Bell Mobil System. 800–331–0500
Tampa	GTE Mobilnet. 800–828–7280
Washington, D.C.	Bell Atlantic Mobile System. 800–255–2355

11

Computers and Data Banks

When Alexander Graham Bell spoke the first sentence to Mr. Watson and word got out that voice could be transmitted over a wire, the world assumed that the days of nonvocal transmission were short-lived. Wrong! With the advent of the computer—and, most important, the personal computer—nonvocal electronic signals are competing for access to telephone lines across the country. Anyone with a computer knows how easy it is to sit before a blinking screen, type a message, and know that as you type your words appear almost instantaneously on another screen across the street or a few thousand miles away. Data transmission, thanks to telephone technology and the facilities of the phone companies, is now available to every household. When people think of computers, IBM is usually the company that comes to mind. However, it was dear old Ma Bell's research at her Bell Laboratories that led to the development of the semiconductor—the genesis of all later transistor technology. When the ax hovered above Ma's head and she had to split off her many operating companies, be assured that Bell Labs were held tight to her bosom. And within months of the divestiture AT&T

announced the availability of both a business and a personal computer in direct competition with IBM.

Although this book is devoted primarily to telephone communications, it would be a serious omission if computers and data banks were not mentioned. For a far more complete overview of computers, check your library for any of the myriad books now available. However, in passing, a few words.

The essential element in computer transmission via telephone is a device called a "modem." A modem is the device that translates your written, numerical, or drawn symbols into electronic impulses, which are then sent to a second modem, which in turn translates the signals back into visual symbols; it is the telephone link between two or more computer terminals. Several years ago a modem looked like a small shoebox attached by wire to a computer. To use it you placed your telephone handset into a special cradle on the modem. Electronic signals generated by the computer were sent over the phone lines to another similar modem anywhere in the world. Now the shoebox has been replaced by one of two devices: a small rectangular box that can be hidden behind your computer, or a printed electronic circuit board that fits easily in the computer itself. With either of these modems dialing a number, waiting for the responding computer signal and disconnecting from the signal can be performed automatically.

In selecting a modem you must choose the speed of transmission, measured in baud rate (the number of bytes, or electronic characters, that can be transmitted over a line), as well as the various gadgets that computer manufacturers like to include to sell more modems. A modem will cost anywhere from $200 to over $600 depending on the various options you select. Be sure to shop around carefully, as the discounts on modems are extremely deep at many computer stores as well as through the various mail-order shops listed in the major computer magazines. One of the largest modem manufacturers, Hayes Microcomputer Products, Inc., lists its Smartmodem 300 at $289, but most mail order dealers take a hundred dollars or more off.

Once you have a modem-equipped computer, you are ready to wander through the world of data bases. Want the latest stock quotations? The local weather report—in Miami? How about the name of a good French restaurant in Chicago? Or if your special field of interest is buying rare books, how about up-to-date data on stolen books? These data, plus a growing assortment of both general and specialized files of information, are now available to the owner of a personal computer. In 1984 more than fourteen hundred different data bases were available to computer users. Certainly, there is nothing secret about the growing sources of data now available to the armchair researcher. In 1983 alone data bases were consulted more than 60 million times. In dollar volume, sales of electronically generated information exceeded $1 billion.

Uniting computers and telephones is something AT&T is very anxious to pursue. For example, a group of two hundred families in Ridgewood, New Jersey, and three hundred fifty households in Palos Verdes and Mission Viejo, California, were part of an experiment conducted by AT&T, the Columbia Broadcasting System (CBS), and the Times-Mirror Company. Each family's television was hooked up with a hand-held control box. By pressing a button the viewer retrieves from a central data file any one of thirty-five hundred pages of news, weather, sports, airline schedules, and the like. The system, called Videotex, is a two-way service. Those families with "the box" can do their banking, order theater or sporting event tickets through Ticketron, play electronic games with other subscribers, send and receive electronic mail, and even exchange favorite recipes.

Although some of these specialized computer data bases will succeed, others will fail to prove commercially viable. In addition to the general, on-line, personal-computer-oriented data bases, there are a myriad of smaller, highly specialized data banks for individuals with special interests. A printed guide, *Directory of On-line Data Bases,* was initiated in 1979 to list the four hundred plus data bases then available. By 1984 the number of data banks had increased to more than thirteen hundred, and the publication is now issued quarterly to stay abreast of the constant changes. Although many of the data bases listed relate to scientific and technical fields, there are more than enough to tempt the nonscientific user. Among the most used specialized data sources are:

LEXIS. The old days of a lawyer spending hours in a law library in search of a specific statute or decision may soon be over. Now a lawyer can use LEXIS to locate and view not just a citation but the full text of a federal court decision, statute, or regulation, as well as other legal references and texts.

WPI. This is a detailed listing of patents issued by the patent offices of the major industrial nations. Each month more than twenty thousand new entries are added to this important data bank.

MEDLINE. Designed primarily for medical professionals in search of specific medical articles that have been published in one or more of the three thousand international journals since 1966, this data base contains well in excess of 5 million references. It is a vital tool for helping many medical professionals diagnose and find treatments for rare diseases.

HOME SUBSCRIPTION SERVICES

By and large the most popular data bases are those that offer their wide range of services to home personal computer owners. Whether you own an inexpensive off-the-shelf computer or a high-priced customized model, access to one of the general data sources is but a modem away.

Anyone can subscribe to a service. Some charge for time plus a hookup fee; others simply charge for your time on-line. All provide a local access number and your own personal password or access key. You are charged on a use basis. This means, obviously, that the longer you remain hooked into the system, the more your monthly bill will be. It's like the phone company: if you're a long talker, you'll pay more. Billing is easy. Most companies just charge your on-line costs to your national credit card, such as VISA, MasterCard, American Express, Diners Club.

The major companies include:

THE SOURCE 1616 Anderson Road, McLean, VA 22102.

800–336–3366.

Founded in 1979 by Source Telecomputing Corporation, the company was bought by the Reader's Digest Association in December 1980 with a major influx of cash supplied in April 1983 by Control Data Corporation. With the editorial and marketing know-how of Reader's Digest combined with the computer and data-base experience of Control Data, it was expected that The Source would soon be the largest and most influential of all consumer-oriented data banks. Those initial expectations have not, as yet, been realized. However, by the summer of 1984 more than sixty thousand members were hooked into The Source's main computers in Langley, Virginia (also the home, coincidentally, of the CIA and its computers).

What do you get for your dollar? The Source offers six main categories of service: Communications, News and Sports, Business and Finance, Travel, Games, and Consumer Services. For example, if you want to fly from Chicago to New Orleans, your on-line service will provide you with the latest schedules plus any changes in fares. Furthermore, The Source has the facility to allow you to make plane or hotel reservations through its travel club.

Costs: To use The Source, there is a one-time membership fee of one hundred dollars with use charges based on the time you spend on-line. The charges vary depending on the speed at which data is transmitted and the time of day you tie into the system. From 7:00 A.M. to 6:00 P.M. on weekdays members pay $20.75 per hour; at all other times the charges are $7.75 per hour, billed in increments of one minute.

You can join The Source at any one of two thousand national computer stores or by calling them directly.

COMPUSERVE 5000 Arlington Centre Blvd. Columbus, OH 43220

800–848–8990

CompuServe has become the nation's largest provider of data to computer users by successfully integrating the use of microcomputers with large data-processing mainframes. Since entering the consumer-oriented videotex industry in 1979, the Columbus, Ohio–based company has seen the ranks of its subscribers surge past the 110,000 mark. In 1980 the company was acquired by H&R Block, Inc. Obviously, CompuServe and The Source are in head-to-head competition. Both provide the same basic service through accessed local telephone numbers. CompuServe, however, has spread its base to include access in more than forty-two foreign countries. Some of CompuServe's capabilities include:

News, Travel & Shopping: Through CompuServe you can have direct access to news from the 128 news bureaus of the Associated Press, and a national bulletin board allowing subscribers to post and review notices of general interest. For-sale advertisements, employment opportunities, requests for information, and announcements of important public meetings are typical postings; the simulated CB allows subscribers to communicate with one another from coast to coast on a real time basis; *The Official Airlines Guide Electronic Edition* contains airline schedules and fares; if you're interested in specific travel information, Travel America is a twenty-four-hour on-line service; weather, sports, and a shopping catalog of over sixty thousand discounted items fill in this broad category of data.

Communications: EMAIL is CompuServe's answer to poor postal service. Although businesses make the most of this service, any subscribers can use it.

Home Management: Aside from the shopping data, CompuServe also offers special data bases for taxes, cash flow, checkbook balancing, insurance, and the like.

Education: Available on-line is a videotex version of Grolier's *Academic American Encyclopedia* plus data on college boards, SATs, and specific information on individual colleges.

Games: There are more than fifty games available on CompuServe. Some of them you play yourself; others involve two or more subscribers. There is even a trivia game to take advantage of a late trend.

Costs: You can buy a start-up kit at any computer store. The kit gives you a password and a local access number plus five free hours of on-line time. The kit sells for $39.00. No other charges are imposed as a subscription fee. Many computer stores throw in the CompuServe starter kit free with the purchase of a modem. On-line costs depend on the speed of transmission. At 300 baud from 8:00 A.M. to 6:00 P.M. the rate is $12.50 per hour; nights and weekends the rate drops to $6.00. At 1,200 baud the daytime rate is $15.00 per hour; nights and weekends the rate is reduced to $12.50 per hour. All rates are charged in one-minute increments.

NEWSNET 945 Haverford Road, Bryn Mawr, PA 19010
800–345–1301

This up-to-the-minute data base is primarily designed to assist business people in obtaining current news on subjects of specialized interest: The service sifts through two hundred newsletters and wire-service transmissions to input data on a wide range of specific topics such as investments, taxes, computers and electronics, publishing, telecommunications, and so forth. To use the service, all you do is check with your own personal computer whenever you want the latest information on your specialized topic, and NewsNet transmits it directly to your screen.

Costs: There is no sign-up fee. You simply pay for the time you are on-line. The basic 300-baud daytime connect charge is $24.00 per hour, billed to the minute. Evening and weekend rates are $18.00 per hour. If you have a 1200-baud modem, just double the price. You dial a local number for access. These charges are for searching or scanning for stories. If you want to read an on-line newsletter, you may be charged more than the base price, depending on the newsletter. For example, if you want to read the current issue of *Real Estate Investing Letter,* you pay $24.00 per hour; choose *ISIS New Electronic Media Newswire* and you'll be charged $120.00 per hour. Most of NewsNet's newsletters, however, are at the lower figure.

DOW JONES NEWS/RETRIEVAL P.O. Box 300, Princeton, NJ 08540
800–257–5114

If you want to lock into a totally business-oriented on-line service, Dow Jones News/Retrieval is one of the largest in the world, with more than 150,000 users. This service provides a wide range of on-line data, including news stories and special features from *The Wall Street Journal* and *Bar-*

ron's, as well as the latest news from the Dow Jones News Service. Users can access specific items through a series of 170 industry and government codes. A text search service allows users to search the full text of *The Wall Street Journal* back to June 1979.

Costs: There is no fee to join the program, and with a standard plan you will not be charged a minimum monthly fee. On-line charges vary depending on the type of service you choose, the baud rate of your modem, and the time of day. If you use a 300-baud modem, you can expect to be charged from $.20 to $.90 per minute in the evenings and on weekends, and from $.90 to $1.20 per minute during the day. With a 1200-baud modem the rate is increased by about 50 percent.

NEXIS 9393 Springboro Pike, Dayton, OH 45401 *513–865–6800*

Mead Data Central, a major information company with headquarters in Ohio and sales offices interspersed throughout the country, offers a wide range of information services for the home computer owner as well as for the operator of a large or small business. In 1983 the company's data products, particularly NEXIS, received a shot in the arm with the acquisition of *The New York Times*'s INFOBANK, which also included the company's The Information Bank—a library of news summaries and abstracts originally created by *The New York Times* Information Service. INFOBANK consists of the Advertising and Marketing Intelligence (AMI) data base, a collection of abstracts from over sixty leading marketing publications prepared by *The New York Times* Company; The Information Bank, a collection of abstracts, including *The Wall Street Journal;* TODAY, a twice-daily news summary that provides general and business news highlights from *The New York Times;* Deadline Data on World Affairs; and the full text of *The New York Times.* NEXIS is an on-line, full-text, general and business news information service. It contains the full text of more than one hundred publications, including a number of the world's leading newspapers, magazines, wire services, and reference materials.

Costs: If you want a Mead Data Central terminal or printer, there is a one-time cost of $200 each plus your monthly rental. There is also a monthly subscription charge of $50 for all users. The price of each search is based on the value and amount of information in the file or group of files being accessed. Prices range from $9 to $18 per search during peak hours. A 50 percent discount applies during off-peak hours, from 7:30 P.M. to 7:30 A.M., local time.

DIALOG INFORMATION SERVICES
3460 Hillview Avenue,
Palo Alto, CA 94304
800–227–1927

One of the largest on-line information retrieval systems, Dialog has been in operation for over a decade, serving thousands of users in business, government, universities, and libraries. More than 100 million items of information are plugged into their computers, including references to books, patents, directories, journals, and newspaper articles. Dialog gives you on-line access to bibliographic citations for all its data plus full text articles and newswires on some of it. For home computer users, Dialog has initiated a new service called *Knowledge Index*. This service is available evenings and weekends and provides personal computer users with convenient low-cost access to a selected group of data bases. Information is divided into eight subjects: Agriculture; Computers/Electronics and Engineering; News and Business Information; Magazines and Books; Medicine and Psychology; Legal; Education; and Government. More than ten thousand international journals are indexed and fed into the Dialog data base, together with federal and state documents, books, reports, newspapers, and so on.

To use Dialog it's best to key your request as specifically as possible. For example, suppose you are interested in learning about available data on malnutrition in Africa. If you ask solely about malnutrition, your computer terminal will be flooded. However, by limiting the search to "Malnutrition"; "Children"; "Nigeria, Chad, Togo"; "1980–present," your search will be easier. You also have your option of a list of citations, or citations plus article abstracts.

There is a problem with Dialog that is not of this fine data bank's making. Once you have a detailed listing of source documents—journal articles, newspapers, government reports, and the like—where do you go to find the complete text of the cited reference? If you're near a major university or large public library, you may find many of the cited journals. However, you may also have to do some calling on your own to locate all the listed sources. Regardless, it is a superb service that will spit out information on practically any subject you can think of.

Costs: There is no subscription or sign-up fee. Your charges are based on three factors: the time you spend online, the data base or data bases you select to use, and whether you choose to print out your data or simply read it off your screen. There are hundreds of data bases from which to select, ranging from ABI/INFORM, at $73.00 per hour ($1.22/min), to Zoological Record, at $78.00 per hour ($1.30/min).

BRS 1200 Route 7, Latham, NY 12110
518–783–1161

BRS offers access to information encompassing the fields of medicine, education, science, health, business, politics, social sciences, energy, and related interdisciplinary areas. The company has more than 50 million records in its file.

Its data bases include *Predicasts Prompt, Books in Print, Ulrich's International, GPO Monthly Catalog, Chemical Abstracts, Harvard Business Review, National Technical Information Service (NTIS), Sociological Abstracts,* among others. Although not as all-encompassing as some of the other banks, BRS does serve a distinct need for those personal computer users who would like access to the specific materials available.

Costs: BRS has several plans. Its "Open Access Plan" provides use of its data base at $35.00 per hour plus royalties (depending on the data base you select) and telecommunications costs. There is no limit to use. The "Subscription Plan" provides you with use of the data base for $16.00 to $30.00 per hour depending on usage. The "Shared Subscription Plan" allows you to share the cost of a subscription with as many as five separate companies or individuals. Each has his or her own password and may be billed separately.

SELECTING YOUR OWN DATA BANK

Obviously, if you feel you have a need for a data bank, be it a general service such as The Source or CompuServe, or a specialized service such as BRS, the first step is to make contact with the service and receive its literature. Any of the above services will be more than glad to flood your mailbox with brochures. After all, they are in business to sell you their wares. Read over the material carefully. Be sure that the service offered is what you need and want. It is very easy to run up a whopping VISA or American Express bill by playing games with a data bank or using a service unnecessarily. After all, it may be convenient to check the flight schedule between New York and Chicago on your computer, but that exercise may cost you five dollars or more. A ten-cent call to your travel agent will give you the identical information. You will probably have to sign an agreement or contract. Read it over carefully. It is a legal instrument and you are bound by its terms. Know what you're signing, and be sure the terms of the agreement conform to your understanding of the service's literature. If you have any questions, don't be afraid to call the company and ask for the information you want. They all have professional salespersons and are used to handling your questions. If a salesperson doesn't have an immediate answer, request that they get back to you.

The chances are you don't need more than one information service. The Source and CompuServe provide a great deal of overlapping data. Each, however, also provides unique data. Make sure you shop and compare those unique services before selecting one of them. The same general rule is true for the more specialized services. Before looking into the various information services, first list your personal or professional needs. Then look for the service that best satisfies those needs.

Finally, give the service of your choice an on-line evaluation using your modem and computer. Does the service operate as advertised, or is it too difficult and confusing? After all, you shouldn't have to be a college professor or computer expert to use a search service. It must suit your goals, needs, and capabilities or it really isn't worth its cost.

— APPENDIX —

Directory of Telephone Services and Equipment

Adam York
Unique Merchandise Mart
Building 6
Hanover, PA 17333
717–637–1600

AT&T Consumer Products
5 Wood Hollow Road
Parsippany, NJ 07054
800–222–3111

Brookstone Company
127 Vose Farm Road
Peterborough, NH 03458
603–924–9511

Cobra Communications
Dynascan Corporation
6460 West Cortland Street
Chicago, IL 60635
312–889–8870

Hammacher Schlemmer
147 East 57th Street
New York, NY 10022
800–368–3584

Hayes Microcomputer Products, Inc.
5923 Peachtree Industrial Boulevard
Norcross, GA 30094
404–449–8791

Krone, Inc.
6950 South Tucson Way
Englewood, CO 80112
800–992–9901

Motorola Inc.
1301 East Algonquin Road
Schaumburg, IL 60172
312–397–1000

Mura Corporation
177 Cantiague Rock Road
Westbury, NY 11590
516–935–3640

Panasonic Company
One Panasonic Way
Secaucus, NJ 07094
201–348–7000

Phone-Mate, Inc.
325 Maple Avenue
Torrance, CA 90503
213–618–9910

Radio Shack
(check your local phone directory)

Sanyo Business Systems Corporation
51 Joseph Street
Moonachie, NJ 07074
201–440–9300

The Sharper Image
680 Davis Street
San Francisco, CA 94111
800–344–4444

Synchronics
Unique Merchandise Mart
Building 42
Hanover, PA 17333
800–621–5800

Teleconcepts, Inc.
36 Holly Drive
Newington, CT 06111
203–666–5666

Uniden Corporation of America
15161 Triton Lane
Huntington Beach, CA 92649
714–898–0558

Webcor Electronics, Inc.
28 South Terminal Drive
Plainview, NY 11803
516–349–0600

The Western Union Telegraph Company
One Lake Street
Upper Saddle River, NJ 07458
800–247–7100

Zoom Telephonics
207 South Street
Boston, MA 02111
617–423–1072

— INDEX —